Changing Climate Politics

Changing Climate Politics

U.S. Policies and Civic Action

Editor

Yael Wolinsky-Nahmias
University of Southern California

Los Angeles | London | New Delhi
Singapore | Washington DC

Los Angeles | London | New Delhi
Singapore | Washington DC

FOR INFORMATION:

CQ Press

An Imprint of SAGE Publications, Inc.

2455 Teller Road

Thousand Oaks, California 91320

E-mail: order@sagepub.com

SAGE Publications Ltd.

1 Oliver's Yard

55 City Road

London EC1Y 1SP

United Kingdom

SAGE Publications India Pvt. Ltd.

B 1/I 1 Mohan Cooperative Industrial Area

Mathura Road, New Delhi 110 044

India

SAGE Publications Asia-Pacific Pte. Ltd.

3 Church Street

#10-04 Samsung Hub

Singapore 049483

Acquisitions Editor: Sarah Calabi

Editorial Assistant: Davia Grant

Production Editor: Stephanie Palermini

Copy Editor: Cate Huisman

Typesetter: C&M Digitals (P) Ltd.

Proofreader: Pam Suwinsky

Indexer: J. Naomi Linzer

Marketing Manager: Erica DeLuca

Cover Design: Jay Mark Johnson

Title of Artwork: PACIFIC COAST HIGHWAY #1

(detail)

Artist: Jay Mark Johnson

Printed in the United States of America

Library of Congress Cataloging-in-Publication Data

Wolinsky-Nahmias, Yael, 1960–

Changing climate politics : U.S. policies and civic action / Yael Wolinsky-Nahmias, University of Southern California.

pages cm.
Includes bibliographical references and index.

ISBN 978-1-4522-3997-2 (pbk. : alk. paper)
ISBN 978-1-4833-1169-2 (web pdf)

1. Climatic changes—Political aspects—United States.
2. Environmental policy—United States. 3. Environmentalism—United States. 4. United States—Environmental conditions.
5. United States—Politics and government. I. Title.

QC903.2.U6W65 2014
363.738′745610973—dc23   2013039536

This book is printed on acid-free paper.

14 15 16 17 18 10 9 8 7 6 5 4 3 2 1

Contents

Preface   vi

Selected Acronyms   xi

About the Contributors   xiii

1. Introduction: Global Climate Politics   1
    Yael Wolinsky-Nahmias

Part I. Changing Climate Policies in the United States   31

2. The Limits of National Climate Policy Making and the Role of the Courts   32
    Kirsten H. Engel

3. A New Era in States' Climate Policies?   55
    Barry G. Rabe

4. Climate Policy Innovation in American Cities   82
    Rachel M. Krause

Part II. Civic Society and Climate Change   109

5. Explaining Public Conflict and Consensus on the Climate   110
    Dennis Chong

6. The US National Climate Change Movement   146
    Robert J. Brulle

7. Environmental Policies on the Ballot   171
    Diana Forster and Daniel A. Smith

8. Consumer Political Action on Climate Change   197
    Lauren Copeland and Eric R. A. N. Smith

9. The Politics of Urgent Transition   218
    Thomas Princen

Index   239

Preface

A RADICAL TRANSFORMATION OF THE RELATIONSHIP between society and nature occurred over the past two centuries. For much of history, the Earth's climate was assumed to be a natural system beyond human control. Early social and political development required people to protect themselves against the unpredictable and capricious forces of nature. Only since the Industrial Revolution, in the wake of a growing world population, economic development, and changing human aspirations, has society acquired the heretofore unimagined capacity to alter the vast oceans, the atmosphere, the seasons of the year, and the climate. With the power to reshape nature, human society is now in a position to affect the fate of future generations, but not always in the most desirable ways.

The idea of a greenhouse effect on the Earth's climate was suggested more than 150 years ago by Irish physicist John Tyndall. Later in the 19th century, Swedish chemist Svante Arrhenius discovered the impact of high levels of atmospheric CO_2 on global temperatures, indicating a possible connection between industrialization, consumption of fossil fuels, and global warming. Scientific research on the role of greenhouse gases (GHG) nonetheless was slow to evolve, as the rise of CO_2 in the atmosphere was considered small and insignificant. This complacency was challenged when Charles David Keeling developed a scientific method for measuring CO_2 in the atmosphere and began recording it systematically in 1959. Keeling's first measurement, at the US Weather Bureau observatory in Mauna Loa, Hawaii, showed an annual mean of 316 parts per million (ppm) atmospheric CO_2. CO_2 levels have risen steadily ever since, and Mauna Loa is now showing measurements of just below 400 ppm, an increase of more than 25 percent in just over a half century. These elevated levels have raised serious concern among scientists, political leaders, and the public about the consequences of higher concentrations of GHG on the Earth's climate and on human society.

Once scientists established the causal connection between GHG and global warming, they focused on three sets of questions: the nature of human influence on climate change, the interaction effects on the climate among different physical and biological systems—the atmosphere, the oceans, and the Earth—and the current and projected consequences of climate change. In addition to gaining deeper knowledge of the physical, chemical, and biological processes involved, researchers have also turned their attention to the societal activities that contribute significantly to global warming. Both world demand for fossil fuels and changing land use are major drivers of climate change.

Yet, much of the research on climate change has concentrated on its scientific dimensions, while many significant social and political aspects of climate change have been understudied. Although climate change is a physical phenomenon, its broad implications require social and political action at every level of society, by individuals, cities, nations, and the international community. The overwhelming evidence of human impact on the climate and the risks of global warming have highlighted the need for a comprehensive climate policy in the United States. However, the formidable economic, technological, and political challenges involved in both mitigation of GHG emissions and adaptation to the changing climate have created a stalemate at the federal level. One consequence of federal inaction has been a decentralized process of climate policy making in the United States that has created more incentives and opportunities for state and city governments and for public participation.

This book explores changing climate politics in the United States by focusing on the relationship between climate change policy and society. It examines progress, innovation, and challenges in climate policies at the federal, state, and local levels, and analyzes how civic society has participated in addressing climate change. The aim of the book is to provide a comprehensive source for examining both policies and civic action on climate change in the United States. By focusing on the social dimension of climate policy, the book emphasizes the role civic society has played in climate policy making and the need for changes in social attitudes and behavior, as well as new policies, to tackle the problems of climate change.

As society begins to address climate change, it is necessary for governments and citizens to engage in a serious dialogue about the challenges and responsibilities they face. The book begins with a discussion of the global dimensions of climate change and the efforts of nations to regulate GHG emissions and apportion costs among countries with varying industrial outputs and economic resources. Despite regular international meetings and the creation of international institutions devoted to climate change, countries have had limited success cooperating on effective global climate policies. Chapter 1 examines how a collective action problem, combined with the high economic costs and long time horizon of climate change effects, have slowed international efforts to address climate change. Similar political and economic challenges have led to stagnation at the US federal level. In Chapter 2, Kirsten H. Engel discusses how federal climate policy making in the United States has gravitated to the courts. Engel examines recent key rulings and their impacts on climate policy. The political impasse at the federal level has also led to a decentralized process of climate policy making and given rise to new avenues for public participation. In Chapter 3, Barry G. Rabe explores innovative approaches by states to address climate change, including California's cap-and-trade policy and key economic and energy measures in other states. Rachel M. Krause, in Chapter 4, discusses the proliferation of policies for the reduction of GHG emissions in American cities, including climate action plans and sustainability measures. The incremental developments of policies at the regional, state, and municipal levels have formed a basis for progress toward addressing climate change in the United States.

The orientations and actions of the public on climate change are examined in the second part of the book. Dennis Chong explores, in Chapter 5, how American public attitudes toward climate change have been influenced by elite politics, scientific information, and political communications. In Chapter 6, Robert J. Brulle analyzes the emergence of the climate change movement in the United States, focusing on the development of climate movement coalitions. Diana Forster and Daniel A. Smith investigate in Chapter 7 the use of direct democracy to pursue environmental policies at the state level, including measures related to climate change. In Chapter 8, Lauren Copeland and Eric R. A. N. Smith explore how citizens have used the market and consumerism to engage in political action on climate change. In Chapter 9, Thomas Princen argues that incrementalism will not suffice to address the problems posed by climate change; instead, he contends that society must reduce its dependence on fossil fuels by delegitimizing their use and should also engage in a transition toward social and political localization.

The contributions to this book suggest that while there is no comprehensive US federal policy on climate change, significant climate policies are emerging through the federal courts and in statehouses and city governments, and social attitudes and behavior are changing. The perspectives in this book represent diverse scholarly fields, including public policy, law, political psychology, political science, and sociology. From these multiple viewpoints, the authors demonstrate the importance of creative policies and strategies, changing attitudes, and public participation.

Acknowledgments

The idea for this book grew out of a seminar entitled Global Climate Change: Policy and Society that I taught at Northwestern University when I was the director of the Program in Environmental Policy and Culture. During a meeting with sales representative Earl Pingel of CQ Press, we remarked on the lack of books on the social and political aspects of climate change. My conviction that addressing climate change requires significant changes in policy, social attitudes, and civic action motivated me to develop this book. My students in that seminar and other courses on environmental politics have been an inspiration to me. It seems courses on environmental issues tend to attract people interested in improving and making a difference in the world.

I was very fortunate to have a group of top experts in the field contribute to this book. I am grateful for their original work and for their intellectual friendship. I especially thank Dennis Chong for his insights and comments on the manuscript and for helping me develop the book. I am grateful for his ability to be both critical and supportive. I also thank Charisse Kiino of CQ Press for her enthusiastic support of this project. Charisse offered excellent advice on the manuscript and kept everything and everyone on schedule. Many thanks to Mark Silberg for his outstanding research support, helpful comments, and assistance in preparing the manuscript. Mark responded to all my requests with enthusiasm and good humor.

Susan Buck, University of North Carolina, Greensboro; Carolyn J. Craig, University of Oregon; Matt Evans, Pennsylvania State University, Altoona; Frank Fischer, Rutgers University–Newark; P. Brian Fisher, College of Charleston; John Freemuth, Boise State University; Marjorie Hershey, Indiana University; Bruce Huber, Dartmouth College; Tim Knievel, Rutgers University; Daniel Press, University of California, Santa Cruz; and Eric R. A. N. Smith, University of California, Santa Barbara provided many important comments and suggestions. I am grateful to all of them. I also thank the Dornsife College of Letters, Arts and Sciences at the University of Southern California for its support. I appreciate the excellent work of Cate Huisman as copyeditor, Stephanie Palermini as production editor, and Davia Grant as editorial assistant at CQ Press.

A note about the cover art: It was a sunny afternoon—aren't they all in Santa Monica?—in March 2013 when Dennis and I stepped into the William Turner Gallery at the Bergamot Station Art Center and first saw the remarkable panoramic images of artist Jay Mark Johnson. Jay's work exploring timeline photography challenges our perceptions of space and time. Imagine taking repeated pictures of the same point in space as the world passes by and splicing them together in a time sequence. Objects that move slowly are elongated, while fast-moving objects are compressed. Things that are stationary appear as straight lines. I am grateful that Jay gave us permission to use one of his photographs and that he also helped to design the cover. The monumental *Pacific Coast Highway #1* beautifully captures, through the use of this photographic technique, the continuous motion on a busy stretch of highway along the shoreline of the Pacific Ocean. More broadly, it offers an artistic exploration and rethinking of the relationship between humans, our machines, and nature, encapsulating in essence what this book is about.

This book was published around a time of big changes in my life when I moved from Chicago to Los Angeles. I thank my parents, Bat-Sheva and Joseph, for their love and encouragement. I thank Orna, Shuki, Avi, Noa, and their families for their friendship and support. I am grateful to my kids, Tamar and Dan, for their endless patience and love. I've been fortunate to have Dennis as a companion in life's many adventures. I thank him for precious support every step of the way.

At a time when climate change has become a reality, critical choices have to be made by governments and individuals. Although the United States is far from achieving a coherent national policy on climate change, important climate policies and social action have emerged. I hope this book offers insights into these developments and that it will stimulate readers to consider the environmental responsibilities we all have.

Yael Wolinsky-Nahmias
Santa Monica, California
October 2013

Selected Acronyms

AB 32	California Assembly Bill 32 (also known as The 2006 Global Warming Solutions Act)
ACEEE	American Council for an Energy-Efficient Economy
ACF	Advocacy Coalition Framework
ANES	American National Election Study
AOSIS	Alliance of Small Island States
AR	Assessment Report
CAP	Climate Action Plan
CARB	California Air Resources Board
CCAP	Chicago Climate Action Plan
CCP	Cities for Climate Protection
CERs	Certified emissions reductions
CDM	Clean Development Mechanism
CFCs	Chlorofluorocarbons
COP	Conference of the Parties
DOE	U.S. Department of Energy
EAP	Environmental Action Plan
ECOS	The Environmental Council of the States
EDF	Environmental Defense Fund
EERS	Energy efficiency resource standard
EMO	Environmental movement organization
EPA	Environmental Protection Agency
EU	European Union
EU-15	The 15 EU member states 1995-2004
EU-25	The 25 EU member states 2004-2006
EU-27	The 27 EU member states 2007-2013
G-77	Group of 77 (developing countries)
GCC	Global Climate Coalition

GDP	Gross Domestic Product
GEF	Global Environment Facility
GHG	Greenhouse gases
ICLEI	International Council for Local Environmental Initiatives
HCFC	Hydrochlorofluorocarbons
IPCC	Intergovernmental Panel on Climate Change
JI	Joint implementation
LA21	Local Agenda 21
MCPA	Mayor's Climate Protection Agreement
MGGRA	Midwestern Greenhouse Gas Reduction Accord
NA 2050	North America 2050: A Partnership for Progress
NAPA	National Adaptation Programmes of Action
NEPA	National Environmental Policy Act
NGO	Non-governmental organization
NOAA	National Atmospheric and Oceanic Agency (U.S.)
NRDC	Natural Resources Defense Council
OECD	Organisation for Economic Co-operation and Development
PCL	Planning and Conservation League
RAN	Rainforest Action Network
RGGI	Regional Greenhouse Gas Initiative
RISA	Regional Integrated Sciences and Assessments
RPS	Renewable portfolio standards
UN	United Nations
UNCED	United Nations Conference on Environment and Development
UNEP	United Nations Environment Programme
UNFCCC	United Nations Framework Convention on Climate Change (1992)
WCI	Western Climate Initiative
WMO	World Meteorological Organization
WTP	Willingness to pay

About the Contributors

Robert J. Brulle is professor of sociology and environmental science in the Department of Culture and Communications at Drexel University. His research focuses on the US environmental movement, critical theory, and public participation in environmental policy making. He is the author of over 50 articles in these areas, and is the author of *Agency, Democracy, and Nature: The U.S. Environmental Movement From a Critical Theory Perspective,* as well as coeditor, with David Pellow, of *Power, Justice, and the Environment.*

Dennis Chong is chair of and professor in the Department of Political Science at the University of Southern California. He studies American national politics and has published extensively on issues of decision making, political psychology, and collective action. He is the author of *Rational Lives,* a study of values, group identification, and conflict over social change. He also wrote *Collective Action and the Civil Rights Movement,* a theoretical study of the dynamics of collective action as well as a substantial study of the American civil rights movement and the local and national politics that surrounded it. Professor Chong's current research on the influence of information and framing in competitive democratic contexts has received several national awards, including the American Political Science Association's Franklin L. Burdette/Pi Sigma Alpha Prize. An active member of the profession, Professor Chong has been elected to the executive council of the APSA, and he is coeditor of *Cambridge Studies in Public Opinion and Political Psychology,* a book series published by Cambridge University Press.

Lauren Copeland is a postdoctoral scholar at the Center for the Environmental Implications of Nanotechnology at the University of California, Santa Barbara. She received her PhD in political science from the University of California, Santa Barbara in 2013. Her research examines how changing citizenship norms, values, and communication technologies affect political behavior and public opinion. She has published articles on these topics in *Political Studies, American Politics Research,* the *Journal of Information Technology & Politics,* and *New Media & Society.*

Kirsten H. Engel is professor of law at the James E. Rogers College of Law at the University of Arizona, where she teaches and researches in the areas of environmental and administrative law with a particular emphasis on federalism and the role of the courts in fostering policy responses to climate change. Engel is the coauthor of an environmental law textbook and numerous book chapters and articles. Her work appears in journals such as the *UCLA Law Review Discourse,* the *Minnesota Law Review,* and the *Ecology Law Quarterly.* Prior to joining the law faculty at the University of Arizona, Engel held appointments within other academic institutions and in the public and nonprofit sectors, including the US Environmental Protection Agency, the Massachusetts attorney general's office, and visiting and permanent professorships at Harvard, Vanderbilt, and Tulane law schools.

Diana Forster is a doctoral candidate in American government at the University of Florida and a research fellow with the Center for Talent Innovation. Her research focuses on the role of religion in American political behavior and on the use of direct democratic policy-making mechanisms within the US states. Her dissertation explores the influence of religious context on political behavior and policy outcomes.

Rachel M. Krause is an assistant professor in the School of Public Affairs and Administration at the University of Kansas. She received her PhD in public affairs from Indiana University in 2011. Her research examines the adoption, implementation, and diffusion of innovative policies by local governments and focuses primarily on municipal sustainability and climate protection initiatives. Her work has appeared in the *Journal of Urban Affairs, Urban Studies, Energy Policy,* and *Review of Policy Research.*

Thomas Princen is associate professor of natural resource and environmental policy at the University of Michigan. He explores issues of social and ecological sustainability, including principles for sustainability, the language and ethics of resource use, and the transition out of fossil fuels. Princen is the author of *Treading Softly: Paths to Ecological Order* (2010/2013), author of *The Logic of Sufficiency* (2005), and lead editor of *Confronting Consumption* (2002). The last two books were awarded the International Studies Association's Harold and Margaret Sprout Award for the "best book in the study of international environmental problems." Princen is also coeditor of *The Localization Reader: Adapting to the Coming Downshift* (MIT Press, 2012), coauthor of *Environmental NGOs in World Politics:*

Linking the Local and the Global (1994), and author of *Intermediaries in International Conflict* (1992/1995).

Barry G. Rabe is the J. Ira and Nicki Harris Chair of Public Policy at the Gerald R. Ford School of Public Policy at the University of Michigan, where he also directs the Center for Local, State, and Urban Policy (CLOSUP). Rabe is also a nonresident senior fellow at the Brookings Institution and a fellow of the National Academy of Public Administration. He has authored or edited two books on climate change policy: *Greenhouse Governance: Addressing Climate Change in America,* and *Statehouse and Greenhouse: The Evolving Politics of American Climate Change Policy.* In 2006 he became the first social scientist to receive a Climate Protection Award from the US Environmental Protection Agency.

Daniel A. Smith is professor of political science at the University of Florida. He has published extensively on direct democracy, campaign financing, and voting rights and elections in the United States. He is the author of *Tax Crusaders and the Politics of Direct Democracy* and the coauthor of *Educated by Initiative,* books examining the politics of ballot initiates in the American states. He frequently serves an expert consultant on election-related lawsuits, and he is the coauthor of a leading college textbook, *State Politics: Institutions and Reform.*

Eric R. A. N. Smith is professor of political science at the University of California, Santa Barbara. He is affiliated with the Bren School of Environmental Science and Management and the Environmental Studies Program at UCSB. Professor Smith's research focuses on environmental politics, public opinion, and elections. He has published several books, including *Energy, the Environment, and Public Opinion,* and numerous articles in journals such as *Political Psychology, Environmental Politics,* and *Political Science Quarterly.*

Yael Wolinsky-Nahmias is an associate professor of the practice of environmental studies and political science at the University of Southern California. She has written on environmental politics, public attitudes toward climate change, and international relations. Wolinsky-Nahmias is the coeditor with Detlef Sprinz of *Models, Numbers, and Cases: Methods for Studying International Relations,* a widely adopted book that examines different approaches to the study of international relations. She previously taught at Northwestern University, where she was the director of the Program in Environmental Policy & Culture.

CHAPTER 1

Introduction

Global Climate Politics

Yael Wolinsky-Nahmias

FOR THE FIRST TIME IN HUMAN HISTORY, in May 2013 the average daily level of CO_2 in the atmosphere reached 400 parts per million (ppm).[1] While 400 ppm is only a symbolic round number, this threshold has not been breached in at least 3 million years, at a time when sea levels were up to 80 feet higher than they are today.[2] Scientists have now clearly established that the rapidly increasing amount of greenhouse gases (GHGs) in the atmosphere is causing significant disruption to the global climate system and poses serious threats to the environment and human society. Abundant evidence of climate change has been discovered in the past few decades, including significant warming of the Earth's surface, especially in the Arctic; melting ice caps; rising sea levels; and increasing ocean temperatures and acidity. In addition, severe droughts in large parts of the United States; ferocious fires in Australia, Russia, and southern Europe; and the flooding of large parts of New York City by Hurricane Sandy in October 2012 have generated much discussion about their causes. While scientists are cautious to link specific weather events to climate change, the unusual magnitude of some of these events is consistent with the predictions of certain climate change models.[3]

The considerable increase in GHG emissions since the Industrial Revolution is largely a direct consequence of energy production, fueled by worldwide transportation and growing consumption. Policy makers, scientists, and majorities of the public in countries around the world now recognize the role of human activity in the accumulation of GHG in the atmosphere.[4] But despite the mounting evidence of climate change and public concern about the issue, effective international action has been slow to develop. International negotiation over climate change began in the late 1980s, leading to the 1992 United Nations Framework Convention on Climate Change (UNFCCC). Although it was signed by more than

150 countries, the UNFCCC and the 1997 Kyoto Protocol that followed have produced limited results. The 1997 Kyoto Protocol in particular failed to achieve some of its central goals, partly because of lack of compliance by countries with the highest GHG emissions levels, including the United States and Canada (both of which later withdrew from the Protocol).[5] Talks to create a new, legally binding[6] agreement to replace the Kyoto Protocol or a second commitment period for the Kyoto Protocol have been slow owing to wide differences in the positions of key parties.

The difficulty of designing effective international climate policy reflects the complexity of the collective action problem involved. A collective action problem arises when individuals (or countries) "acting out of pure self-interest, are unable to coordinate their efforts to produce . . . certain public goods they find desirable."[7] Mitigating climate change is a public good that requires the cooperation of many countries, including the largest current and future emitters of GHGs. However, many countries prefer to avoid the costs of cutting emissions and to increase their energy consumption in pursuit of industrial development. Moreover, the largest emitters are often not the countries that will suffer the greatest impact of climate change, and much of the damage will occur well into the future. Achieving international cooperation on climate policy is thus complicated by the highly unequal shares of GHG emissions among countries, the disproportionate distribution of risks, and the temporal structure of the issue.

While some elements of the collective action problem of international climate policy have persisted, important changes have occurred since the signing of the 1992 UNFCCC. Climate change is a multidimensional problem, and, the focus of negotiations has shifted over time, as countries have updated their priorities on different aspects of the problem. The strategic calculations of countries have also been affected by movements in their relative positions as contributors to climate change or as nations vulnerable to its consequences. In addition, the information conditions have improved, as newer models of climate change are able to provide more consistent and detailed projections. Finally, the negotiating coalitions have changed recently, raising new concerns about growing gaps within and among different blocs. The influence of these developments on international climate policy making is complex. Some changes, for example, improved information, have created more urgency and thus increased the pressure on policy makers to compromise, while other changes, such as rapidly increasing emissions levels by developing countries, have caused greater difficulty in achieving international cooperation.

In this chapter, I discuss international efforts to address climate change by examining key aspects of the collective action problem. My intent is to provide an international context for the analysis of climate change politics in the United States. In many ways, the successes and failures of international cooperation on climate change have provided the impetus for emerging policies and civic action on the issue in the United States.

The Emergence of the International Climate Regime

During the past few decades, scientists have acquired a deep understanding of the causes and effects of global climate change. Beginning in the late 1970s, the World Meteorological Organization and the United Nations Environment Programme organized conferences that discussed the serious threats posed by increasing concentrations of GHG in the atmosphere. Since then, more advanced climate change modeling has provided progressively more reliable projections of the time frame of climate change and its varying effects on different parts of the world. Four major Assessment Reports (ARs), and the first part of the Fifth Assessment Report (Working Group I) by the Intergovernmental Panel on Climate Change (IPCC) have presented clear evidence of significant warming, rising sea levels, more frequent extreme weather events, and increased susceptibility of poor populations.[8] But despite this considerable body of scientific knowledge, international negotiation over how to address climate change has shown relatively little progress.

First Steps, the 1992 UN Framework Convention on Climate Change

When the international community began to acknowledge the need for climate policy in the late 1980s and early 1990s, there was little relevant policy experience to guide the assessment of policy alternatives. The most similar international policy problem, which had emerged about a decade earlier, was the depletion of the ozone layer caused mainly by the release of chlorofluorocarbons (CFCs) into the atmosphere through use of aerosol sprays and other human activities. As an atmospheric issue, new to scientists, the public, and decision makers, the discovery of the "ozone hole" was a uniquely global challenge. Yet the international community was able to address it successfully within a few years. Following a general declaration on the need to address the issue, many countries adopted a ban on CFCs.[9] When concern about GHGs evolved in the 1980s and 1990s, the international community tried to follow a similar diplomatic path, by first creating a declarative framework convention followed by an institutional structure for negotiating specific reductions. However, important differences between these two problems, in the scope of their causes and the availability of low-cost alternatives, have led to different results. The story of the climate change regime has thus been much more complicated than that of the ozone layer regime, and serious conflicts of interest have continued to hamper international cooperation.

A key challenge in developing effective climate change policy has been finding a fair solution to an externality, created mostly by a small number of developed countries, with widespread implications for all. Designing a fair and sustainable solution is further complicated by escalating levels of GHG emissions in some developing countries. With these considerations in mind, early negotiation on climate policy dealt with the guiding principles for an international agreement on emissions reductions, issues of fairness, and the market's role in providing incentives to control emissions and foster adaptation.

Initially, international discussions on climate policy seemed promising. The 1988 Toronto conference, the first international conference to discuss climate policy, was organized by scientists, policy makers, and environmentalists and sponsored by the Canadian government. The conference set the first specific (nonbinding) target for carbon emissions reductions—the "Toronto target"—calling for countries to reduce their carbon emissions by 20 percent below 1988 levels by 2005. Following the Toronto conference, the European Union decided to take action to stabilize GHG emissions at 1990 levels by the year 2000.

The First Assessment Report (AR1) by the IPCC in 1990 raised serious concerns about the threat of a changing climate. The report led to the first milestone in international climate policy—the UNFCCC, signed at the 1992 United Nations Conference on Environment and Development in Rio de Janeiro, also known as the Earth Summit. The Convention laid out key principles and goals for the international community as well as an institutional framework for future climate negotiations. The signing of the UNFCCC by 154 developing and developed countries signified both broad concern and optimism that the international community could work together to protect the planet and human society. Thousands of representatives from 225 countries, including 144 heads of state and more than 2,400 representatives of nongovernmental organizations (NGOs), participated in what was one of the largest international environmental conferences to date. Parallel to the Earth Summit, about 17,000 people attended the NGO "Global Forum." Such broad participation set the tone for a unique international policy-making process that featured scores of scientists and mobilized countless ordinary citizens, who became involved through the growing number of NGOs working on climate change issues.

The 1992 UNFCCC established a conceptual framework and key principles for international cooperation on climate change. The Framework Convention's main objective was to create a broad political commitment for "stabilization of greenhouse gas concentrations in the atmosphere at a level that would prevent dangerous anthropogenic interference with the climate system."[10] The Convention established several key principles, including common but differentiated responsibilities—recognizing the obligation of all countries to address climate change but also the huge differences among them (especially in their capacities and contributions to the problem) as a basis for requiring industrialized countries to make significant emissions cuts.

Even at the time of the 1992 talks, countries were already concerned that action would be delayed due to incomplete scientific understanding of climate change. Consequently, Article 3 of the UNFCCC stated what later became known as the precautionary principle: "When there are credible threats of serious and irreversible damage, lack of full scientific certainty shall not be used as a reason for postponing cost-effective measures to prevent environmental degradation."[11] The remainder of the Convention dealt with two main issues—specific commitments by the parties, and the establishment of institutions and procedures for continued cooperation. Specific obligations by the parties included collecting and reporting annual inventories of GHG emissions and preparing for climate adaptation. The Convention established the Conference of the Parties (COP), the key institutional mechanism for continued negotiation and coordination. The COP has met every year since 1995 to negotiate further agreements and to discuss the implementation of countries' commitments.

Similar to the Vienna Convention on Protection of the Ozone Layer and to the 1979 Convention on Long-Range Transboundary Air Pollution, the UNFCCC established the basis for the international climate change regime. An important limitation of the Convention is that it did not contain legally binding emissions cuts, which was a concession to US demands. Thus, signing the UNFCCC did not create a legal commitment by countries to reduce their emissions. It also did not define what constitutes a dangerous level of emissions or of temperature change, two questions that would become highly controversial in subsequent talks. However, with broad participation and wide recognition of the inequities involved and the need to achieve a fair solution, hopes were high that meaningful international climate policy would follow.

Progress, the 1997 Kyoto Protocol

European countries took a leading role in efforts to advance a binding treaty for the reduction of GHG emissions following the 1992 Framework Convention. The first important step was the Berlin Mandate, adopted by the first COP in 1995, for negotiating a legally binding protocol for the reduction of GHG emissions. Agreement on the Berlin Mandate occurred around the same time as the release of the IPCC Second Assessment Report (AR2), which stated that human activities were having a "discernible" impact on climate, and that global inaction would have serious consequences for the planet.

Following two years of complex negotiation, national delegations convened in Kyoto in 1997 to finalize and sign an agreement to reduce emissions of the six most dangerous GHGs. The participation of the largest emitter—the United States— remained uncertain until the last moment. American negotiators, led by Vice President Al Gore, argued that the United States accepted the notion of unequal responsibilities in reducing emissions, but that developing countries nonetheless should participate and agree to cut their emissions.[12] It was only after the United States modified this position that the parties to the Kyoto Protocol adopted the principle of common but differentiated responsibilities and implemented it in a set of binding commitments for emissions reductions. Thirty-seven industrialized countries and the European Union ("Annex I Parties") committed to an average reduction of 5.2 percent in their emissions relative to their 1990 levels by 2008–12.[13] By setting varying target reductions in industrialized countries, while allowing developing ("non-Annex 1") countries to increase their emissions, the Kyoto Protocol introduced what seemed at the time to be an innovative policy solution for addressing the vast inequalities in countries' responsibilities for altering the Earth's climate.

The Protocol allowed each country to choose specific methods for complying with its emissions reduction commitments. During the negotiation, the United States successfully advanced "flexible mechanisms," including three types of measures: international emissions trading, the clean development mechanism (CDM), and joint implementation (JI). The operation and employment of those measures were finally agreed on four years later in the 2001 COP-7 in Marrakesh, Morocco.

In the beginning, the Kyoto Protocol was viewed as a major step toward global reduction of GHG emissions. As time progressed, however, many in the scientific community grew increasingly skeptical that the target cuts prescribed by the Protocol would be sufficient to address climate change. In addition, President George W. Bush's decision to withdraw from the Kyoto Protocol shortly after taking office in 2001 dealt a serious blow to the UNFCCC process. While the United States had played a limited leadership role in international climate negotiation, the decision of the most powerful country in the world and the largest emitter of GHGs at the time to withdraw from the Kyoto process raised serious concerns about the future of international climate policy. US withdrawal also prevented the Kyoto Protocol from entering into force until 2005, when political maneuvering by the European Union facilitated Russia's ratification of the Protocol.

Despite the ratification of the Kyoto Protocol in 2005, compliance by some Annex I participants has been limited. Several countries, including the largest emitter among Annex I countries, the United States (which never ratified the Protocol and later withdrew its signature), have not met their goals. However, the EU-15 has surpassed its first Kyoto commitment of cutting GHG emissions by 8 percent below 1990 levels. Table 1.1 compares CO_2 emission trends of the top 10 global emitters. It shows the significant gap between China and the U.S. and all other top emitters. The table highlights the dramatic increase in CO_2 emissions by China, India, Indonesia, and South Korea, all non-Annex 1 countries that do not have commitment targets within the Kyoto Protocol. Indeed, China and India combined have now reached a level of emissions similar to that of the OECD countries combined. Only two of the top ten emitters, Russia and Germany, have reduced their CO_2 emissions below 1990 levels (Germany's emissions decreased by about 20 percent between 1990 and 2012, even though its GDP per capita doubled). Table 1.1 also

TABLE I.I   Comparison of CO_2 Emission Trends

Country (top 10 global emitters 2012*)	% of world total emissions 2012* (% 1990)	Per capita emissions 2012* (metric ton CO2/ capita)	Kyoto Ratification Year	Kyoto Target (2008–2012) % reduction from 1990 levels	% change in emissions from 1990 to 2012* **
China	28.6 (11.4)	7.1	2002	N/A***	**275.8**
United States	15.1 (22.6)	16.4	—	−7	4.1
India	5.7 (3.0)	1.6	2002	N/A***	**198.1**
Russian Federation	5.1 (11.1)	12.4	2004	0	−27.3
Japan	3.8 (5.3)	10.4	2002	−6	14.1
Germany	2.3 (4.6)	9.7	2002	−8	−20.9
South Korea	1.8 (1.1)	13.0	2002	N/A***	**154.2**
Canada	1.6 (2.0)	16.0	2002, then withdrawn 2012	−6	24.2
Indonesia	1.4 (0.7)	2.0	2004	N/A***	**209.0**
Mexico	1.4 (1.3)	4.0	2000	N/A***	57.3

* 2012 data is based on The European Union Emissions Database for Global Atmospheric Research (EDGAR) estimates.

** Over 100% increase since Kyoto is noted in boldface. Below Kyoto target change is shaded in gray.

*** N/A- Not applicable, non-Annex 1 countries.

Sources: The Emissions Database for Global Atmospheric Research (EDGAR): edgar.jrc.ec.europa.eu; United Nations Framework Convention on Climate Change: http://unfccc.int/resource/docs/publications/08_unfccc_kp_ref_manual.pdf, and http://unfccc.int/kyoto_protocol/status_of_ratification/items/2613.php

demonstrates the significant gaps in per capita emissions among the largest emitters, as the U.S. and Canadian rates are almost two times greater than China's and ten times greater than India's per capita emissions. When all regulated GHG are considered, the gaps are even larger, with the U.S. and Canada having twice the per capita rate of Germany, and almost three times the rate of China.

Although total world CO_2 emissions increased by about 50 percent between 1990 and 2012, the Kyoto Protocol did have important contributions. The Protocol created a formula, accepted by the vast majority of countries, for implementing the principles of the UNFCCC. The Protocol thus translated the key moral principle of common but differentiated responsibilities into a workable policy framework. In addition, the Protocol, and the UNFCCC more generally, provided specific institutional arrangements and a set of goals for the development of a market mechanism for GHG emissions cuts. Finally, the Protocol helped to create a political dynamic of commitment that influenced climate policy making in many countries. Thus, while the Kyoto target reductions may not have been ambitious enough, and even as such have not been reached by some countries, the Kyoto Protocol represented significant progress in developing the moral framework for climate action, the technical aspects of the implementation, and domestic policy making in some countries.

The imminent expiration of the Kyoto Protocol, combined with continued US inaction, and the significant increase in world GHG emissions, required a reassessment of the next stage in international climate policy in the mid to late 2000s. The 13th session of the COP convened in Bali, Indonesia, in 2007 to design "long-term cooperative action, now, up to and beyond 2012, in order to reach an agreed outcome and adopt a decision at its fifteenth session [in Copenhagen]."[14] European countries took the lead in designing the "Bali Road Map" as it became known, which tried to recreate a shared vision for addressing mitigation, adaptation, technology, and financing. COP-13 in Bali also established working groups to devise new emissions reductions commitments for industrialized countries after the 2012 expiration of the Kyoto Protocol.

Copenhagen and Beyond

As the highly anticipated 2009 COP-15 in Copenhagen approached, demands for action on climate change were growing, including a call for action by UN Secretary General Ban Ki-moon. A few months prior to the COP-15 in Copenhagen, the G-8 countries agreed on the goal of limiting global warming to 2 degrees Celsius (3.6 degrees Fahrenheit) by 2100. However, the rift between developed and developing nations continued to widen, with China and India demanding that industrialized nations commit to a 40 percent reduction in their CO_2 emissions over the next decade. China and India also signed a collaboration pact to increase energy efficiency and the use of renewable energy and to invest in the development of cleaner coal technologies. As China and India probably intended through this move, the pact by the two largest emitters of CO_2 in the developing world on the eve of the Copenhagen conference made it more difficult for developed countries to challenge developing nations to commit to specific target reductions.

In December 2009, more than 40,000 people, including leaders from 193 countries, scientists, and members of nongovernmental organizations, gathered in Copenhagen to design an aggressive and innovative set of policies to address climate change. Expectations were high for a legally binding emissions treaty to replace the Kyoto Protocol, which was set to expire in 2012. Archbishop Desmond Tutu summed up the attitudes of those calling for serious action:

> Science has spoken on the urgent need to tackle the challenge of climate change. Now it is time to listen to our consciences. There is a clear moral imperative to tackle the causes of global warming. Worldwide, we have the chance to start turning the tide of climate change, but only if all governments commit themselves to a fair, binding and sustainable climate agreement in Copenhagen.[15]

The negotiating text for COP-15 shared this sentiment, requiring deep cuts in emissions and calling for an "economic transition" toward lower-emission lifestyles, with developed countries showing "leadership in mitigation commitments or actions, [and] in supporting developing country Parties in undertaking adaptation measures . . . [by] assisting them through the transfer of technology and financial resources."[16] There was also considerable emphasis on adaptation issues in the preparatory documents, including recommendations for designing flood-proof housing in vulnerable areas, warning systems for climate-related disasters, and an insurance framework for climate change–affected infrastructure. However, top US officials tried to lower expectations before the start of the COP, and as the Conference progressed, hopes for a binding climate treaty quickly faded. Last-minute negotiation between President Obama and the leaders of China, India, Brazil, and South Africa yielded the Copenhagen Accord, a nonbinding political agreement, which was signed by only 28 of the 193 participating countries. Subsequently, more than 100 countries added their signatures.

The Copenhagen Accord is a concise three-page document dealing with critical issues for the future of humanity and the planet. It addresses the need to slow the rate of temperature increases and cut GHG emissions, emphasizes expanding use of forests as "sinks" for absorbing CO_2, and provides guidelines on adaptation and assistance to developing countries. The Accord contains a number of positive developments, calling for the continuation of the Kyoto Protocol beyond 2012 and declaring, for the first time in the history of international climate policy, a specific goal—limiting the increase in average global temperature to less than 2 degrees

Celsius above preindustrial temperature levels. The developed countries also agreed to raise $30 billion by 2012, and an additional $100 billion *per year* by 2020 for a new Green Climate Fund aimed primarily at funding adaptation measures in developing nations and a transition to low-carbon economic development. The greatest weakness of the Copenhagen Accord is that it is not legally binding. Since countries can thus change their pledges and positions at any time, the Accord offers a less stable arrangement and creates doubts about the likelihood of broad implementation of any costly commitment. International compliance is by choice, and indeed, by mid-2013, less than $12 billion had been delivered to the adaptation fund.

The failure by scores of world leaders in Copenhagen to design a binding agreement to replace the expiring Kyoto Protocol drew heavy criticism. James Hansen, then director of NASA's Goddard Institute for Space Studies and a renowned climate scientist, stated that "the developed nations want to continue basically business as usual so they are expected to purchase indulgences to give small amounts of money to developing countries."[17]

Continued negotiations by subsequent COPs produced some progress toward extending the Kyoto Protocol. COP-17 in Durban, South Africa, held in 2011, agreed to establish a second legally binding protocol for GHG reductions by all countries. The conference also decided on the features of the Green Climate Fund, including simplified access to the funds and greater transparency. The 2012 COP-18 in Doha, Qatar, reaffirmed the extension of the Kyoto Protocol through a second commitment period for Annex I parties, to be finalized by 2015 and implemented by 2020. During the second commitment period, parties will reduce GHG emissions by at least 18 percent below 1990 levels between 2013 and 2020. However the composition of the parties in the second commitment period is different from that in the original Annex I. Notably absent are the United States, Canada, Japan, and Russia, who together account for more than 25 percent of world CO_2 emissions. The countries that joined the second commitment period, including all the EU members, Australia, and Ukraine, account for less than 15% of world CO_2 emissions.[18] In November 2013, COP-19 in Warsaw, Poland, agreed on the submission process for countries' proposed target reductions, and established an international mechanism for "Loss and Damage" to provide greater protection to the most vulnerable populations against the adverse effects of climate change. These commitments were achieved only after some high drama, as 132 member nations walked out in protest, demonstrating the deep divisions on central issues including protection for vulnerable countries and mitigation commitments by developing countries.

A key aspect of a future protocol, if the Durban platform were maintained, would be the requirement that developing countries also commit to specific, differentiated GHG emissions reductions. There is wide public agreement in many developing societies that their country has "a responsibility to deal with climate change." Public acceptance of the idea that one's own country has an obligation to take action averages around 90 percent in developing countries—higher on average than public support for this notion in developed countries.[19] Yet, few developing countries governments have followed public sentiment on this issue. Most G-77 + China group members have continued to object to shifting too much of the burden of GHG emissions cuts to large, developing countries.

In June 2012, all UNFCCC member countries, along with thousands of scientists and representatives of nongovernmental organizations and the private sector, gathered in Rio de Janeiro for the 20th anniversary of the 1992 Earth Summit. Rio+20 sought to reaffirm international commitment to sustainable development, green economy, and the eradication of poverty. This was also the 20th anniversary of the UNFCCC, which was signed at the 1992 Earth Summit. But to the dismay of many, climate change was not one of the official themes of the Rio+20 conference. The issues of sustainable development and green economy are closely related, both conceptually and practically, to climate change policy, so there was a certain symbolism in the absence of the subject of climate change per se from the formal agenda of Rio+20. The conference resolutions expressed "profound alarm" and a "grave concern" for rapidly increasing emissions and failed policies; however, many felt this was a missed opportunity to obtain international commitment to urgently needed climate change action.[20]

International Climate Policy Making as a Collective Action Problem

As a global problem that will seriously affect human society for generations to come, climate change provides an opportunity for countries to unite in an effort to overcome the dangers it presents. Indeed, for almost a quarter of a century, virtually all the countries of the world have been engaged in a complex, multifaceted process of climate change policy making. However, efforts to design climate policy present a classic problem of collective action. No single country can eliminate the risk of dangerous GHGs in the atmosphere, and most countries believe they have an incentive to free ride on other countries' efforts to reduce their emissions. The collective action problem is further complicated by the considerable variation across countries in their contribution to the problem and their vulnerability to the impacts of climate change. Furthermore, the countries responsible for the highest emissions are not the same countries that are the most threatened by climate change. In addition, most of the worst consequences are expected to occur well into the future. Finally, a substantial global improvement in emissions levels requires a significant commitment of resources by a large number of countries.

More broadly, there are four key dimensions of the collective action problem in international climate policy making. First, there is the range of issues—economic, political, and technological—implicated in climate change negotiations. Second, there is asymmetry among participants in their contribution to the problem and vulnerability to its consequences. Third, there are asymmetries of information across countries. Fourth, there is instability of the negotiating blocs. In the following section I examine how each of these issues has changed over time, and discuss the impact of these changes on the specific challenges involved in collective action for climate mitigation and adaptation.

Multiple Issue Set and the Shift Toward Adaptation

Designing comprehensive climate policy requires serious consideration of multiple issues, including limits on GHG emissions, adaptation to climate change, technology transfers, and financial mechanisms. At the start of the negotiation for the 1992 UNFCCC, it was clear countries had different interests and priorities regarding these issues. Much of the emphasis in the early stages focused on designing a broad, legally binding agreement for reducing GHG emissions. This remained a key goal of the parties leading to the Kyoto Protocol and more recently to its extension and second commitment period. However, efforts to achieve substantial cuts in carbon emissions have had limited success. A recognition that the currently high levels of GHGs in the atmosphere will remain there for at least 100 years, even if world emissions are reduced in the near future, has produced growing alarm and greater attention to adaptation policies at both the international and state levels.

The UNFCCC established specific financial arrangements through the Global Environment Facility to fund adaptation projects in developing countries, based on the understanding that wealthier nations have a responsibility to assist poorer countries. However, available funds have been far from adequate, and most countries have been slow to develop specific adaptation plans. While the predicted cost of global adaptation varies widely, a 2010 World Bank study estimates an annual cost of $75–100 billion between 2010 and 2050, with a significant portion of that cost earmarked to assist developing countries.[21]

In addition to consistent projections in the IPCC Assessment Reports of high impact on coastal population centers, many countries have begun to experience damages linked to the changing climate. Floods and droughts are endangering communities, affecting critical resources such as water and agriculture, and changing entire ecosystems. Local and state governments have consequently increased their efforts to design adaptation policies. This tendency has contributed to the evolving decentralization of climate change policy, especially in the United States. From coastal management to changes in agricultural practices, state and local governments are trying to address their vulnerabilities more independently. However, lower-level adaptation "always takes place within the constraints and opportunities engendered by antecedent collective action and collective inaction."[22]

While some benefits of adaptation are clearly local, the goal of protecting human society and nature from the impact of climate change has the features of a public good. International climate change policy is a multi-issue problem that encompasses the provision of two related but distinct public goods. The first public good is protecting the atmosphere and restoring its capacity to protect life on Earth. To achieve this goal, countries must agree to substantial, legally binding mitigation commitments. The second public good is protecting human life and nature from the potentially devastating impact of climate change, through adaptation.

Most of the early effort in international climate negotiation was devoted to protecting the Earth and society by reducing GHG emissions levels. The limited success of mitigation, combined with the discernible impact of climate change, has motivated a simultaneous effort to protect people, infrastructure, and ecosystems through adaptation. COP-16 (2010) in Cancun adopted the Cancun Adaptation Framework "to enhance action on adaptation, including through international cooperation and coherent consideration of matters relating to adaptation under the Convention."[23] Because effective adaptation requires substantial funding, it potentially offers wealthy, high-emitter countries a choice between contributing to the public good of climate mitigation or assisting highly vulnerable countries to adapt. Such assistance is essential, but it does not address the core problem of excessive global GHG emissions. The idea that nations can choose between paying for adaptation and mitigation is misleading, because lack of action on mitigation will only produce growing adaptation costs.[24] Nonetheless, debates over mitigation and adaption characterized the 2009 COP-15 in Copenhagen, where in the absence of agreement on mitigation, wealthy major-emitter countries agreed on a large adaptation fund to assist highly vulnerable developing countries.

Following the UNFCCC provisions, 50 countries including all 48 least developed countries (as of November 2013) have developed National Adaptation Programmes of Action (NAPAs). These plans are just a first step for identifying the most critical needs for climate adaptation and are, in most cases, far from implementation. Furthermore, of the most vulnerable poor countries, only a handful have created and submitted such plans to the UNFCCC secretariat.[25]

While both mitigation and adaptation are crucial for protecting life on Earth, it remains to be seen whether the increased emphasis on necessary adaptation, with its funding mechanism, will allow some countries to dodge perhaps even more economically and politically costly GHG emissions cuts.

Asymmetry

The urgency of developing adaptation policies has focused attention on the extreme asymmetry among nations in both their vulnerability to climate change and their capacity to adapt. Asymmetry has also characterized the history and current state of world GHG emissions, with developed countries being responsible for the majority of GHG emissions over time. The starkest asymmetry is actually between the top emitters and the smallest contributors to world emissions. China, the largest emitter of CO_2 since 2007, is responsible for about 29 percent of total world emissions of CO_2. The top three emitters—China, the United States, and India—are responsible for about 50 percent of total CO_2 emissions, while each of the 90 lowest-emitter countries contributes less than one hundredth of one percent of total world emissions.[26]

While these key imbalances among countries continue, important changes have been occurring in the global order. Top emitters among developing countries, that is, China and India, are now emitting almost as much CO_2 as all OECD countries combined. This change suggests that the application of the common but differentiated responsibility principle may need revision, as many developing countries pursue economic development with little regard to carbon emissions. In particular, China has been facing growing pressure to reduce its emissions. Although China has declared specific goals to reduce its carbon intensity (the amount of CO_2 emitted per unit GDP) by 40–45 percent by 2020, compared with 2005, it maintains its right to continue to develop and increase emissions.[27]

Asymmetries also characterize the projected consequences of climate change, with some northern countries potentially standing to benefit from global warming, while small island nations are unlikely to survive the rising seas in their current locations. Geographic location is a key factor behind the wide variation in vulnerability, with Southern Hemisphere countries expected to face more serious challenges. While there remains some uncertainty regarding how much the sea level will rise and the time horizon over which this will occur, developing countries with high population concentrations in coastal areas are extremely vulnerable. Overall, as discussed in the IPCC Assessment Reports, many of the most vulnerable countries are poorer developing countries that have limited capacity to design and implement adaptation measures. In addition to small island countries, some of the other developing countries most susceptible to climate change include Bangladesh, Vietnam, and Egypt.[28]

One area in which the asymmetry in vulnerability is having profound consequences is the impact of global warming on agriculture. Developing countries are predominantly located in lower (warmer) latitudes and are highly dependent on local and subsistence agriculture. A comparative analysis of the effects of global warming on agriculture shows that, even taking into account a possible small positive impact of CO_2 through enhanced photosynthesis in some crops, continued warming will result in developing countries suffering much higher losses in agricultural productivity than developed countries (losses as high as 10–25 percent).[29]

However, the level of asymmetry in costs and benefits across nations might actually be reduced by the secondary costs of climate change. Lower agricultural production could lead to famine and destabilize political regimes in some countries.

High variation in precipitation (even in areas that are not especially arid) or increased frequency of extreme weather events can wreak havoc on the production of future commodities and affect world prices. Rising sea levels may spawn population movement that could affect political stability and security in some parts of the world. If climate change significantly impairs agricultural production, or if sea level rise inundates highly populated coastal areas, many countries may face indirect consequences and costs, including industrialized Northern Hemisphere countries such as Russia that may otherwise be less vulnerable to climate impact. Such potentially high secondary costs of climate change highlight the interdependent vulnerability of the international community.

The changes in asymmetry in both causes and impacts of climate change have called for a serious discussion of the moral aspects of climate change policy. The common but differentiated responsibilities principle has guided international climate policy, including the emissions reduction commitments of the Kyoto Protocol and its extension. The Protocol allowed developing countries to pursue economic development and avoid emissions target reductions. With China and India increasing their GHG emissions well beyond the emissions of all (in the case of China) or most (in the case of India) developed countries, the interpretation of the principle has been called into question.

Over time, three different approaches have emerged. According to the original interpretation, all countries are responsible for addressing the problem; developed nations, however, should be legally obligated to cut their emissions due to their historical and current large contributions. A different interpretation of the principle calls for all large emitters, regardless of the history of a country's contribution to total emissions and its current development level, to pursue substantial emissions reductions. A third interpretation suggests the principle should apply within societies as well, based on different socioeconomic classes. Thus, upper-middle-class citizens in China whose consumption surpasses that of some poorer populations in the developed world should be held responsible and required to change their behavior.[30]

The first two interpretations represent contrasting political and moral approaches to international climate policy. The first approach addresses the highly unequal historical contribution of different nations to the problem, while the second approach emphasizes current levels and likely future trends in GHG emissions. The third approach disregards national boundaries by highlighting socioeconomic variations in consumption that may serve as the criteria for the distribution of costs. While much international climate policy making has focused on the rationale and justification of country-level policies, the expansion of local climate policies (discussed in Chapter 4) and the implementation of market mechanisms (discussed in Chapter 3) have broadened the discussion on fairness and feasibility.

Information Conditions

A central problem in designing climate policy has been the availability and reliability of information about causes and effects of climate change. Key questions during the early stages of international negotiation revolved around the relative contribution of different gases and sectors to the problem, and the contributions of different countries. More recently, efforts to improve information have focused on the rate of climate change under varying conditions, predicted regional and local impacts, and the cost-effectiveness of alternative policies. Scientific and policy research on climate change has expanded dramatically in recent years, providing better comprehension of causality, interaction effects, predicted rates of impact, and policy alternatives. However, limited and complex information has often slowed down international climate policy making. One of the main challenges in constructing a clear enough picture for policy making has been the need to review, assess, and synthesize mounting research from different scientific fields, including atmospheric sciences, oceanography, geology, chemistry, and biology. In this sense, climate change has posed an unprecedented information problem.

During the early stages of climate negotiation, countries had variable access to scientific information and professional expertise on climate change. However, the growing role of the IPCC and of nongovernmental organizations helped small countries overcome some information problems. The IPCC, established in 1988 by the World Meteorological Organization and the United Nations Environment Programme, brings together thousands of scientists who specialize in different fields related to the climate system from around the world. The IPCC reviews, assesses, and synthesizes hundreds of scientific studies on climate change to make the information more accessible for policy making and public release. Over the years the IPCC has become the most authoritative source of climate change science, and it has played a central role in the international climate regime. Furthermore, the IPCC has had an important role in fostering an international consensus on the role of human activities in changing the Earth's climate. Although the IPCC is an intergovernmental organization, its involvement, and of scientists more generally, in international climate policy has been criticized by some as too far reaching and influential and by others as not assertive enough. Overall, climate scientists have provided invaluable information about the science of climate change and reduction levels of CO_2 that may be necessary to avoid catastrophic warming of the planet.

The IPCC has completed four major Assessment Reports (1990, 1995, 2001, 2007). The first part of the IPCC Fifth Assessment Report was released in 2013 (and the synthesis report is expected to be released in October 2014). The reports are based on scientific studies and are reviewed and compiled by hundreds of scientists from around the world. The Assessment Reports have raised serious concerns regarding threats of climate change. Under the "business as usual" scenario, climate change is predicted to lead to a global average temperature rise of 1.8–2.2 degrees Celsius (almost 4 degrees Fahrenheit) by 2081–2100, according to the 2013 medium estimates.[31] This would be the highest sustained temperature change over the last 10,000 years, and it is estimated to be well above a tipping point, or a threshold, that would set in motion irreversible shifts in the climate system. In addition, climate change could trigger dangerous, abrupt, nonlinear responses of the climate system, such as rapid deglaciation of polar ice sheets, which are hard to predict. Climate change is also likely to have a very broad range of impacts, including warmer surface temperatures, more frequent heat waves, more severe storms including hurricanes and tropical cyclones, rising sea levels, increased intensity of both floods and droughts, loss of farming productivity, spread of infectious diseases, and extinction of species and loss of biodiversity.

The IPCC Assessment Reports have provided a fairly consistent set of predictions over time regarding climate change. In particular, two key projections have been at the center of attention—the rises in average global temperature and in sea levels. Table 1.2 presents key summary projections from the first five Assessment

TABLE 1.2   IPCC Assessment Reports Projections

IPCC Assessment Report (AR) and publication year	Projected range of temperature rise by 2100* (degrees Celsius)	Projected global mean sea level rise by 2100* (meters)
First AR, 1990	1.5–4.5	1.00
Second AR, 1995	1.0–3.5	0.15–0.95
Third AR, 2001	1.4–5.8	0.1–0.88
Fourth AR, 2007	1.1–6.4**	0.18–0.59***
Fifth AR, 2013 (Working Group I part)	1.1–4.8^	0.26–0.82

* Over a range of estimated scenarios, and compared with 1990 levels for AR1-3. For AR4, projection ranges are for 2090–2099 relative to 1980–1999, and for AR5 ranges are for 2081-2100 relative to 1986-2005.

** The best estimates by the six emissions scenarios range from 1.8 to 4.0.

*** Projection does not include future feedback effects and rapid, dynamic changes in ice flow.

^ Except for the lowest scenario

Sources: The Intergovernmental Panel on Climate Change, First Assessment Report (1990), Second Assessment Report (1995), Third Assessment Report (2001), Fourth Assessment Report (2007), Fifth Assessment Report (Working Group I, 2013).

Reports. Scientists can provide only a range of projections, rather than a specific projection regarding both average global temperature and sea levels, because of the internal natural variability in the climate system (e.g., El Niño) and the broad range of possible emissions (40–110 percent increase between 2000 and 2030). The uncertainties and upper ranges for temperature rise are larger than those for sea level rise mainly because of possible carbon–climate feedback effects. The physical science portion of the Fifth Assessment Report declares that "warming of the climate system is unequivocal, and since the 1950s, many of the observed changes are unprecedented over decades to millennia."[32] The report shows global surface temperature change for the end of the 21st century is likely to exceed 1.5 degrees Celsius relative to 1850–1900 for all scenarios except one, and is likely to exceed 2 degrees Celsius for several scenarios. In addition, the report declares: "It is virtually certain that there will be more frequent hot and fewer cold temperature extremes."[33] The report's predictions about temperature rise were produced by a new generation of complex climate models that takes advantage of computational capacity that is far superior to what was available when the Fourth Assessment Report was released.

While the direction of scientific predictions has been consistent, the information conditions of international climate negotiation have markedly changed since the 1992 UNFCCC was signed in Rio de Janeiro, Brazil. Most important, scientific analysis of the complex processes of climate change has greatly improved, and more complete information is available about the range of consequences. Newer models of climate change integrate more physical processes (e.g. land, ice sheets) and improved representation of other factors such as clouds. The quality of scientific information about climate change is considerably higher, and there has been greater exploration of interconnections and feedbacks across the atmosphere and the oceans. And there are more models today, offering increased reliability and consistency of projections on important aspects of climate change such as changes in precipitation.[34]

However, uncertainty about precise consequences is likely to continue due to the difficulty in predicting future emissions levels, natural variability in the climate system, and the response of the climate system to much higher levels of GHGs (termed *climate sensitivity*).[35] In particular, the impact of climate–carbon cycle feedbacks, such as the reduced terrestrial and ocean capacity to absorb CO_2 due to warming, and the full effects of changes in ice sheet flow, are especially difficult to integrate into climate models.[36] The lack of clear scientific information on the dangerous threshold of climate change may have a negative impact on the negotiation process, reduce nations' incentives to cooperate and diminish the possibility of a cooperative solution to the problem.[37]

Future models are expected to provide more reliable information on regional and local impact of climate change. The National Atmospheric and Oceanic

Agency (NOAA) currently supports 11 Regional Integrated Sciences and Assessments (RISA) in the United States, primarily in the south and southwestern parts of the country. More specific information about regional and local effects of climate change is valuable to communities trying to protect themselves from climate impact using a variety of adaptation measures.

Much more information has also become available regarding alternative mitigation policies, especially cap and trade and carbon tax. The broadest international experimentation with cap and trade has been the European Union system, created in 2005. While errors in the design of the program resulted in less than satisfactory results, analysts and policy makers have learned important lessons about key aspects, such as pricing and emission permits. The Regional Greenhouse Gas Initiative (RGGI) of nine northeastern US states provides a fairly successful example of a limited scale cap and trade system focused on reducing CO_2 emissions only from the power sector. Barry G. Rabe discusses these and other market-based mitigation policies in Chapter 3.

Overall, greater transparency of scientific information, especially through the IPCC Assessment Reports, has significantly reduced the level of uncertainty among policy makers and the public about the basic processes of climate change and its consequences. Paradoxically, improved information about high levels of threat due to climate change has not led to the implementation of the precautionary principle by key countries such as the United States. However, greater specificity about areas of risk may have strengthened moral arguments by highly vulnerable countries and nongovernmental organizations, and could increase the priority given to adaptation policies and assistance.

International Participation and Negotiating Blocs

Ever since the negotiation over the 1992 UNFCCC, countries have tried to ensure wide participation in international climate policy making. Indeed more than 190 countries participated in the negotiation over the 1992 Framework Convention, and virtually all countries take part in the climate regime, including the annual COPs. In addition, hundreds of nongovernmental organizations have participated in climate change politics. Participation in the process has not been costless, as countries have had to develop and assess their emissions portfolios and are required to continue to report on a variety of implementation issues. However, as in other instances of international cooperation, participation carries benefits, even for the least powerful, in the form of having a voice and at times being able to affect outcomes, or at least prevent one's least desirable outcomes.

An interesting aspect of the international effort to address climate change has been the coalition structure that formed early to help in the negotiation process. The Kyoto Protocol's Annex I and Annex II countries define two broad coalitions in the international climate regime. But the high number of participating countries (194) and the differences in goals and priorities within each of these two groups led to the evolution of a more elaborate coalition structure. Most countries participate in climate negotiation as part of long-standing international coalitions such as the G-77 + China or the European Union. All parties are organized into five regional groups, such as the Asian States and the Latin American and Caribbean States, following the tradition of the United Nations. Other negotiating groups were formed for the purpose of promoting their specific interests regarding climate change. For example, AOSIS—the Alliance of Small Island States—includes 43 countries, some of which are already under existential threat due to rising sea levels. The United States, Canada, Australia, Japan, and Russia formed the Umbrella Group that promotes minimizing the economic impact of emission reductions through the use of market mechanisms. Most of the party groupings have been pretty stable until recently, though new blocs have emerged, such as the Environmental Integrity Group, which was formed in 2000 and now includes Mexico, Liechtenstein, Monaco, the Republic of Korea, and Switzerland.

For successful outcomes, negotiating coalitions must keep up with the interests and priorities of all members.[38] However, shifting priorities and interests of some countries due to changes in their own or other countries' levels of emissions and strategic considerations have made it difficult to maintain consensus in some of the negotiating groups. Prior to the 2009 COP-15 in Copenhagen, divisions in the G-77+ China group widened. The largest and most diverse group among the negotiating blocs, the G-77 + China group is an ongoing coalition of 133 developing countries with broad experience in international negotiation. Member countries have been united in their demand for emissions cuts by developed countries. However, G-77 + China countries vary significantly in their specific goals and strategies, contribution to climate change, level of vulnerability, and adaptation capacity. These differences took center stage during the Copenhagen COP-15 crisis, when it became clear that a legally binding agreement for replacing the Kyoto Protocol was out of reach. China, India, Brazil, and South Africa then signed a formal agreement, on November 28, 2009, creating a separate negotiating bloc— the BASIC group of four of the largest developing countries—united by their support for a second commitment period under the Kyoto Protocol (and by the lack of binding commitments for target reductions). Extending the Kyoto Protocol without changing the principle distinction between Annex I and non-Annex 1 countries will obviously release BASIC countries from having to make legal commitments to reduce their emissions.

The Copenhagen negotiation failed to establish a new binding protocol for GHG reductions because of the deep differences of views among countries and the lack of US willingness to commit to target reductions. The increasing divisions in the G-77 + China bloc, including the BASIC and the Latin American and

Caribbean groups acting separately, weakened G-77 + China as a bloc, but it also highlighted the strong support of leading developing countries for the second commitment period strategy.

AOSIS also had difficulty in maintaining a collective stand in Copenhagen. The 43 member countries face similar threat from sea level rise but vary markedly in their level of economic development. Although Singapore has one of the highest GDPs per capita in the world, it preferred to align itself with the much poorer small island countries rather than join other industrialized countries. During COP-15 in Copenhagen, AOSIS countries were divided on whether any agreement, including one that has no binding GHG emissions reductions commitments, is preferable to no agreement at all. At the end, most AOSIS states did support the Copenhagen Accord, yet the internal conflict weakened their position during the negotiation process.

Overall, the voluntary grouping of the 194 UNFCCC member countries into negotiating blocs has increased efficiency in communication, reduced transaction costs, and made the multifaceted negotiation process more manageable. At times, it has helped some of the least powerful parties, for example, AOSIS, to amplify their voices, as long as they were able to present a unified position. In other cases, the negotiating blocs have consolidated their members' interests and may have sharpened the contrast between the goals of different blocs, thereby making successful agreements tougher to achieve.

Conclusion

Climate change remains an extraordinarily challenging policy issue owing to the collective action problems it presents, its scientific complexity, its long-term consequences, and the high cost of policy measures. Since the issue emerged in the late 1980s, important changes have occurred in international climate policy making. The recent focus on adaptation issues, in addition to mitigation, highlights the asymmetry in vulnerability and capacity among countries and has elevated the financial needs of poor, climate-vulnerable countries to the top of the international climate negotiation agenda. Consequently, wealthy high-emitter countries could gain even more negotiating latitude to substitute monetary transfers for fundamental changes in their economies and way of life. Regional and local approaches to adaptation have demonstrated the potential for multilevel climate policy making while posing new demands for reliable scientific projections about the impact of climate change.

Asymmetry remains a key feature of the collective action problem of addressing climate change. However, when the secondary costs of climate impact are taken into consideration, cross-national variation in vulnerability is reduced, as fewer countries stand to gain from climate change and the overall level of vulnerability is higher. This broader view of interdependent vulnerability should help facilitate international cooperation. Changes in the information conditions, especially higher transparency, greater reliability, and more elaborate information about the risks of inaction, could help to create a more positive domestic policy environment. Yet, as suggested by Dennis Chong in this volume, more changes are necessary among political and business elites, especially in the United States, to instigate a profound change in public attitudes and policy.

Finally, internal instability of some of the negotiating blocs may yield mixed results for the negotiation. By separating themselves from the G-77 + China group, the largest emitters among developing countries are now freer to make stronger demands on developed countries. However, they may also be exposing themselves to greater pressure to undertake emissions reduction commitments separately from the broader G-77 + China group.

To be successful, international climate change policy must resolve some of the collective action problems discussed here. Some scholars have suggested that greater emphasis on the low-probability catastrophic consequences of climate change is key to motivating countries to discount asymmetries and focus on urgent measures.[39] Indeed critics suggest the IPCC should stress the possibility of catastrophic impact on the Earth more than the low probability of insignificant impact. The IPCC has instead followed scientific criteria for reporting the range of possible scenarios. This approach might change as the time horizon for the impact of climate change is shortened.

Policy makers must consider the secondary costs of climate change, and the economic and political benefits of pursuing low-carbon emissions development paths, in both developed and developing nations. The combination of greater attention to the moral implications of inaction, changing attitudes among social and business elites, and continued economic success in advanced economies that have significantly reduced GHG emissions offers the clearest model for implementing the serious changes required for a safe, prosperous, global society and a healthy planet.

Plan for the Book

International climate policy making is critical for global progress on the issue of climate change. However, it is inherently slow and may not provide the necessary response to the threats of a changing climate. During the past decade, the need for alternative approaches has become more urgent. As the largest emitter of GHG gases in history, the United States holds a special responsibility for addressing climate change. While political stalemate at the national level has slowed federal responses, it has led to a decentralized process of climate policy making and created unique opportunities for public participation and democratic responsiveness.

This book discusses progress and innovation in climate policies at the federal, state, and local levels and the individual and collective efforts being made by citizens to address climate change. Examining both government action and citizen participation allows us to integrate the institutional and social dimensions of climate change problems and solutions, and to critically analyze the operation of democratic processes on this urgent domestic and international issue. Collectively, we offer a balanced assessment of the progress and challenges of addressing climate change in the United States.

Following the review of international climate change policy in this chapter, Part I of the book evaluates the role of the federal government, the courts, statehouses, and cities in tackling climate change. Part II studies the growing role of civic society in climate action plans, exploring public opinion, the US climate movement, policy making through ballot measures, consumer action, and the prospects of a social transition toward a more sustainable society.

Kirsten H. Engel's chapter, "The Limits of National Climate Policy Making and the Role of the Courts," considers the lack of a US national climate policy and the consequences of shifting the conflict over climate policy to the federal courts. Engel argues that because of this shift, federal climate policy is largely a product of legal frameworks rather than of long-term environmental and political planning. The redirection of climate policy making to federal courts has profoundly affected both the substance and the success of federal climate policy. The chapter demonstrates that the process of developing federal policy through the courts has meant that key aspects of climate policy making are either not being addressed or are being dictated by existing statutory authorities. Engel points out that the limited federal climate policy today is largely a product of executive branch actions originally triggered by court mandates.

Barry G. Rabe, in "A New Era in States' Climate Policies?", examines the central role of the states in the development of American climate policy. The chapter reviews recent state policy experimentation and innovation, focusing on major areas of state climate policy making, including the Environmental Protection Agency's 2012 decision to extend Clean Air Act provisions to greenhouse gases, the leading role of California in climate policy, and other single-state and regional policy initiatives. The unexpected discovery of massive deposits of natural gas throughout the United States has triggered a major shift in the fuel used to generate electricity, from coal to natural gas. The chapter evaluates the role of states in the debate over fossil fuels and concludes with an assessment of the influence and consequences of state policies on US emissions trends.

Over the past two decades, US local governments, nonprofits, and community-based organizations have initiated and led hundreds of climate action plans. In her chapter on "Climate Policy Innovation in American Cities," Rachel M. Krause examines these measures, looking at their historical background and current status.

The chapter offers an overview of different types of local climate policies and discusses the motivations leading to their emergence. Krause explores key case studies, including Chicago's climate action plan, and assesses how some communities have approached the challenge of climate mitigation and adaptation. The chapter concludes with a discussion of the continuing challenges involved in addressing climate change at the local level.

The discussion of local climate policy sets the stage for the analysis of public attitudes, civic organizations, and collective choices in the second part of the book. In his chapter titled "Explaining Public Conflict and Consensus on the Climate," Dennis Chong examines how the American public has become ideologically polarized on the issue of climate change and how this division might be bridged in the future. Chong shows that despite growing scientific evidence and increasing public awareness, climate change continues to be a partisan issue in the United States, with federal climate policy viewed as a trade-off between economic and environmental interests. The chapter explores why the messages of the scientific community on the causes and consequences of climate change have not more effectively reduced this partisan divide to produce greater agreement on climate policies. Chong argues that the division in public opinion has its origin in competing elite interests, and he concludes that achieving a more general public consensus requires both greater elite leadership on the issue and creative policies that reconcile the competing values and economic interests underlying partisan and ideological polarization.

As climate change has increasingly become the new focus of environmental concern, a social movement advocating for the mitigation of GHG emissions has emerged in the United States. In Chapter 6, "The US National Climate Change Movement," Robert J. Brulle provides a detailed description of the national climate change movement, discussing its historical development from 1980 to the present. In just 30 years, the movement had grown from a few groups into a major component of the US environmental movement, consisting of more than 250 unique organizations. Many of these groups participate in coalitions that focus on climate change action. Brulle examines the organizational makeup of these coalitions and their cultural and political influence. The chapter highlights the emergence of dominant institutions within the climate change movement as well as the diversity of issues, approaches, and influence of different alliances within the movement.

Diana Forster and Daniel A. Smith, in "Environmental Policies on the Ballot," study state environmental ballot measures from 1904 to 2010. Ballot measures provide opportunities for environmental activists to advance climate policies through public campaigns. The authors offer a historical overview of ballot measures at the state level and examine the success rate of different categories of environmental measures (e.g. land, energy). Forster and Smith then employ statistical analysis to identify sources of success across the 349 proenvironment ballot measures. The chapter examines several case studies, including California's 2008 Proposition 10—the California Alternative Fuels Initiative—and Maine's 2010 Question 2—the Energy Efficiency Bonds Issue. The authors conclude that despite significant variation in success rates across different categories of proenvironment measures, direct democracy at the ballot box offers environmental activists important means to achieve their goals.

As public awareness of the threat of climate change continues to grow, environmental groups and individuals have tried to create economic pressure through consumer choices to advance their goals. In Chapter 8, "Consumer Political Action on Climate Change," Lauren Copeland and Eric R. A. N. Smith discuss the rise of political consumerism on climate-related issues, examining the use of two key strategies—boycotting and buycotting of products and services. The authors explore how consumers have approached transportation and energy issues to assess individual and group efforts to pressure businesses. The chapter evaluates the effectiveness of political consumerism in leading businesses to create more environmentally friendly products, and more generally in achieving social and political change. Copeland and Smith argue that especially in the absence of a comprehensive national climate policy, political consumerism offers an effective tool to promote more sustainable practices in the marketplace.

The last chapter, "The Politics of Urgent Transition," by Thomas Princen, offers a dramatic reframing of the issue of climate change. Princen maintains that the future of human society depends on whether people can make a transition to a fundamentally different state of affairs. Rather than an incremental change of policies and individual behavior, the chapter argues that climate change and other environmental crises require both a major shift in politics and a social reorganization toward localization (with regional, national, and international dimensions). The chapter discusses the biophysical necessity and the moral obligation to reconstruct key aspects of society, including delegitimization of fossil fuels, in order to sustain human society and the planet as we know it.

The multidisciplinary nature of problems caused by climate change requires consideration of broad, integrated solutions that include the participation of all elements of society, from government institutions to civic organizations and individual citizens. Bringing together experts from different areas, including public policy, law, political psychology, political science, and sociology, this book offers a critical analysis of one of the most profound national and international challenges of our time.

The presidential election of Barack Obama for a second term created high expectations in the United States and around the world for a new era in American climate change policy. Although the United States still lacks a comprehensive national climate policy, recent federal regulations, court rulings, state and local policies, and public participation are creating a significant set of changes that have the potential to lead us toward a more secure future.

Suggested Readings

Chasek, Pamela S., and Lynn M. Wagner, eds. *The Roads From Rio: Lessons Learned From Twenty Years of Multilateral Environmental Negotiations.* New York and London: Routledge, 2012.

Harrison, Kathryn, and Lisa Sundstrom, eds. *Global Commons, Domestic Decisions: The Comparative Politics of Climate Change.* Cambridge, MA: MIT Press, 2010.

Hoffman, Matthew J. *Climate Governance at the Crossroads.* Toronto: University of Toronto Press, 2011.

References

Adger, Neil, Jouni Paavola, and Saleemul Huq, eds. *Fairness in Adaptation to Climate Change.* Cambridge, MA: MIT Press, 2006.

Andonova, Liliana B., Michele M. Betstill, and Harriet Bulkeley. "Transnational Climate Governance." *Global Environmental Politics* 9, no. 2 (2009): 52–73.

Bodansky, Daniel. "The Durban Platform: Issues and Options for a 2015 Agreement." Arlington, VA: Center for Energy and Climate Solutions, 2012. http://www.c2es.org/docUploads/durban-platform-issues-and-options.pdf.

Chasek, Pamela S., and Lynn M. Wagner, eds. *The Roads From Rio: Lessons Learned From Twenty Years of Multilateral Environmental Negotiations.* New York and London: Routledge, 2012.

Christoff, Peter. "Cold Climate in Copenhagen: China and the United States at COP15." *Environmental Politics* 19, no. 4 (2010): 637–656.

Cole, Daniel H. *Climate Change and Collective Action.* Working Paper, Workshop in Political Theory and Policy Analysis, Indiana University, Bloomington, IN, 2007.

Fariborz, Zelli, and Harro van Asselt. "The Institutional Fragmentation of Global Environmental Governance: Causes, Consequences, and Responses." *Global Environmental Politics* 13, no. 3 (2013): 1–13.

Giddens, Anthony. *The Politics of Climate Change.* Malden, MA: Polity Press, 2009.

Haita, Corina. *The State of Compliance in the Kyoto Protocol.* Venice: International Center for Climate Governance, 2012.

Harrison, Kathryn, and Lisa Sundstrom, eds. *Global Commons, Domestic Decisions: The Comparative Politics of Climate Change.* Cambridge, MA: MIT Press, 2010.

Hoffman, Matthew J. *Climate Governance at the Crossroads.* Toronto: University of Toronto Press, 2011.

Hovi, Jon, Detlef F. Sprinz, and Arild Underdal. "Implementing Long-Term Climate Policy: Time Inconsistency, Domestic Politics, International Anarchy." *Global Environmental Politics* 9, no. 3 (2009): 20–39.

Intergovernmental Panel on Climate Change (IPCC). *Climate Change: The IPCC Scientific Assessment* [First Assessment Report]. Geneva: IPCC, 1990. http://www.ipcc.ch.

Intergovernmental Panel on Climate Change (IPCC). *IPCC Second Assessment: Climate Change 1995* [Second Assessment Report]. Geneva: IPCC, 1995. http://www.ipcc.ch.

Intergovernmental Panel on Climate Change (IPCC). *Climate Change 2001: Synthesis Report* [Third Assessment Report]. Geneva: IPCC, 2001. http://www.ipcc.ch.

Intergovernmental Panel on Climate Change (IPCC). *Climate Change 2007: Synthesis Report* [Fourth Assessment Report]. Geneva: IPCC, 2007. http://www.ipcc.ch.

Intergovernmental Panel on Climate Change (IPCC). *Climate Change 2013 The Physical Science Basis* (Working Group I) [Fifth Assessment Report]. Geneva: IPCC, 2013. http://www.ipcc.ch.

Kim, So Young, and Yael Wolinsky-Nahmias. "Crossnational Public Opinion on Climate Change: The Effects of Affluence and Vulnerability." *Global Environmental Politics* 14, no. 1 (2014): 79–106.

Knutti, Reto, and Jan Sedlacek. "Robustness and Uncertainties in the New CMIP5 Climate Model Projections." *Nature Climate Change* 3 (2013): 369–373, doi:10.1038/nclimate1716.

Min, Seung-Ki, Xuebin Zhang, Francis W. Zwiers, and Gabriele C. Hegerl. "Human Contribution to More-Intense Precipitation Extremes." *Nature* 470 (2011): 378–381.

Moore, Frances, C. "Negotiating Adaptation: Norm Selection and Hybridization in International Climate Negotiations." *Global Environmental Politics* 12, no. 4 (2012): 30–48.

Oberthür, Sebastian, and Olav Schram Stokke, eds. *Managing Institutional Complexity: Regime Interplay and Global Environmental Change*. Cambridge, MA: MIT Press, 2011.

Pielke, Roger, Jr., Gwyn Prins, Steve Rayner, and Daniel Sarewitz. "Lifting the Taboo on Adaptation." *Nature* 445, no. 7128 (2007): 597–598.

Roberts, Timmons, and Bradley Park. *A Climate of Injustice: Global Inequality, North-South Politics, and Climate Policy*. Cambridge, MA: MIT Press, 2007.

Tamirisa, Natalia. "Climate Change and the Economy." *Finance and Development* 45, no. 1 (March 2008). http://www.imf.org/external/pubs/ft/fandd/2008/03/tamirisa.htm.

Tol, Richard S. J., Thomas E. Downing, Onno J. Kuik, and Joel B. Smith. "Distributional Aspects of Climate Change Impacts." *Global Environmental Change* 14, no. 3 (2004): 259–272.

Victor, David G. *Global Warming Gridlock: Creating More Effective Strategies for Protecting the Planet*. Cambridge, UK: Cambridge University Press, 2011.

Wagner, Lynn M., Reem Hajjar, and Asheline Appleton. "Global Alliances to Strange Bedfellows: The Ebb and Flow of Negotiating Coalitions." In *The Roads From Rio: Lessons Learned From Twenty Years of Multilateral Environmental Negotiations*, edited by Pamela S. Chasek and Lynn M. Wagner, 85–106. New York and London: Routledge, 2012.

Weaver, Christopher P., Robert J. Lempert, Casey Brown, John A. Hall, David Revell, and Daniel Sarewitz. "Improving the Contribution of Climate Model Information to Decision Making: The Value and Demands of Robust Decision Frameworks." *Climate Change* 4, no. 1 (2013): 39–60.

Wolinsky-Nahmias, Yael, and So Young Kim. "Public Concern, Knowledge, and Support for Costly Policies and Actions for Climate Change." Paper presented at the 32nd Annual Scientific Meeting of the International Society of Political Psychology, Dublin, July 2009.

Notes

*I am grateful to Mark Silberg for his superb research assistance. I also thank Charisse Kiino and Detlef Sprinz for their excellent comments, and Dennis Chong for many helpful conversations and valuable improvements of the chapter.

1. A single reading of slightly over 400 ppm was recorded at a NOAA's observatory station in Barrow, Alaska, in May 2012. But the 2013 400-ppm reading at Mauna Loa, Hawaii, is part of a historical recording in that station, which produced the famous "Keeling Curve," the longest continuous record of atmospheric CO_2 going back to 1958.

2. Ron Prinn, "400 ppm CO_2? Add Other GHGs, and It's Equivalent to 478 ppm," Oceans at MIT Featured Stories, June 6, 2013, http://oceans.mit.edu/featured-stories.

3. Seung-Ki Min, Xuebin Zhang, Francis W. Zwiers, and Gabriele C. Hegerl, "Human Contribution to More-Intense Precipitation Extremes."

4. Yael Wolinsky-Nahmias and So Young Kim, "Public Concern, Knowledge, and Support for Costly Policies and Actions for Climate Change," paper presented at the 32nd Annual Scientific Meeting of the International Society of Political Psychology, Dublin, July 2009.

5. European countries that failed to achieve their Kyoto targets include Greece, Portugal, and Spain. For more information, see International Energy Agency, *CO₂ Emissions from Fuel Combustion* (Paris: International Energy Agency, 2012), http://www.iea.org/c02highlights/c02highlights.pdf.

6. A legally binding international agreement creates the highest form of state commitment, because it is typically ratified through domestic legislation by signatory countries. That process makes it legally binding under each country's domestic law as well.

7. Dennis Chong, *Collective Action and the Civil Rights Movement* (Chicago: University of Chicago Press, 1991), 5.

8. The rest of the fifth report is scheduled for publication in 2014.

9. Though some concern about thinning of the ozone layer in certain areas during particular times of the year continues.

10. Article 2 of the *United Nations Framework Convention on Climate Change* (1992), http://unfccc.int/resource/docs/convkp/conveng.pdf.

11. Ibid.

12. This position reflected Senate Resolution 98, known as the Byrd-Hagel Resolution, regarding the conditions for the United States becoming a signatory to any international agreement on greenhouse gas emissions under the UNFCCC.

13. Reduction targets vary from 5 to 8 percent for most countries, while Australia, Iceland, and Norway were allowed to increase their emissions by 8, 10, and 1 percent respectively, and New Zealand, Russia, and Ukraine were allowed to keep their emissions at the 1990 levels.

14. "Decision1/Bali Action Plan," in *Report of the Conference of the Parties on Its Thirteenth Session,* Held in Bali from 3 to 15 December 2007, 3–7, http://unfccc.int/resource/docs/2007/cop13/eng/06a01.pdf.

15. Desmond Tutu, in "International Calls for Action," http://unfccc.int/meetings/copenhagen_dec_2009/items/5067.php. I thank Mark Silberg for this quote.

16. *Negotiating Text.* Bonn: UNFCCC Ad Hoc Working Group on Long-Term Cooperative Action Under the Convention, 2009.

17. James Bone, "Climate Scientist James Hansen Hopes Summit Will Fail," *Times Online,* December 3, 2009, http://www.thetimes.co.uk/tto/environment/article2144839.ece.

18. 2011 data, The Emissions Database for Global Atmospheric Research (EDGAR), http://edgar.jrc.ec.europa.eu.

19. So Young Kim and Yael Wolinsky-Nahmias, "Crossnational Public Opinion on Climate Change: The Effects of Affluence and Vulnerability," *Global Environmental Politics* 14, no. 1 (2014).

20. Jennifer Morgan, "Rio 20 in the Rear View: A Missed Opportunity for Climate Change Action," *WRI Insights,* June 29, 2012, http://insights.wri.org/news/2012/06/ri020-rear-view-missed-opportunity-climate-change-action.

21. World Bank, *The Cost to Developing Countries of Adapting to Climate Change* (Washington, DC: The World Bank Group, 2010).

22. Neil Adger, Jouni Paavola, and Saleemul Huq, *Fairness in Adaptation to Climate Change* (Cambridge, MA: MIT Press, 2006), 7.

23. UNFCCC, *Cancun Adaptation Framework*, 2013, http://unfccc.int/adaptation/items/5852.php.

24. Mitigation and adaptation costs may be interrelated, as unabated emissions create a long-term pressure on ever-increasing adaptation costs, while high adaptation costs may increase pressure to reduce emissions. I thank Detlef Sprinz for suggesting this point.

25. UNFCCC, NAPAs received by the Secretariat, accessed December 12, 2013, https://unfccc.int/adaptation/workstreams/national_adaptation_programmes_of_action/items/4585.php.

26. Calculated from data included in The Emissions Database for Global Atmospheric Research (EDGAR), "Trends in Global CO_2 Emissions 2012 Report," http://edgar.jrc.ec.europa.eu/CO2REPORT2012.pdf; and United Nations Statistics Division, carbon dioxide emissions (excluding LULUCF), http://mdgs.un.org/unsd/mdg/SeriesDetail.aspx?srid=749&crid=.

27. Xinhua News, "China Announces Targets on Carbon Emission Cuts," November 26, 2009, http://news.xinhuanet.com/english/2009-11/26/content_12544181.htm.

28. Susmita Dasgupta, Benoit Laplante, Craig Meisner, David Wheeler, and Jianping Yan, "The Impact of Sea Level Rise on Developing Countries: A Comparative Analysis," World Bank Policy Research Working Paper 4136, February 2007.

29. William Cline, *Global Warming and Agriculture: Impact Estimates by Country* (Washington, DC: Peterson Institute for International Economics, 2007).

30. Juhua Fan, "Cosmopolitanism and the Principle of Common but Differentiated Responsibilities," unpublished manuscript, 2013.

31. Intergovernmental Panel on Climate Change (IPCC), *Climate Change 2013 The Physical Science Basis* [Fifth Assessment Report] (Geneva: IPCC, 2013), http://www.ipcc.ch.

32. Intergovernmental Panel on Climate Change (IPCC), *Climate Change 2013: Summary for Policy Makers, Working Group I* [Fifth Assessment Report] (Geneva: IPCC, 2013), http://www.ipcc.ch.

33. Ibid.

34. Phillip B. Duffy, "Uncertainty in Future Climate: Learn to Live With It," presentation at the University of Southern California Schwarzenegger Institute, Los Angeles, California, April 8, 2013.

35. Ibid.

36. On these and other elements of climate sensitivity, see more in Reto Knutti and Jan Sedlacek, "Robustness and Uncertainties in the New CMIP5 Climate Model Projections," doi:10.1038\'rr x 54354444454545444/nclimate1716.

37. Scott Barrett and Astrid Dannenberg, "Climate Negotiations Under Scientific Uncertainty," *Proceedings of the National Academy of Sciences of the United States* 109, no. 43 (2012): 17372–17376.

38. Lynn M. Wagner, Reem Hajjar, and Asheline Appleton, "Global Alliances to Strange Bedfellows: The Ebb and Flow of Negotiating Coalitions," in *The Roads From Rio: Lessons Learned From Twenty Years of Multilateral Environmental Negotiations,* edited by Pamela S. Chasek and Lynn M. Wagner (New York and London: Routledge, 2012), 85–106.

39. Daniel H. Cole, "Climate Change and Collective Action," Working Paper, Workshop in Political Theory and Policy Analysis, Indiana University, Bloomington, IN, 2007.

PART I

Changing Climate
Policies in the United States

The Limits of National Climate Policy Making and the Role of the Courts

Kirsten H. Engel

Introduction

The tripartite nature of the national government in the United States, together with the nation's federal structure, provides multiple avenues for developing policy responses to a broad spectrum of economic, social, and environmental problems. Lack of action by one branch of the national government does not necessarily mean the other two will be likewise inactive. Similarly, lack of action by one level of government—the national government, for instance—does not guarantee silence at the state or local levels of government. The institutional powers and constraints of each branch and level of government may prompt different policy-relevant responses. Of particular relevance to this chapter, a lack of action at the national level can create coalitions and opportunities for action at the state and local levels of government and vice versa.

Few contemporary issues better illustrate these truisms than climate change. Multiple pathways can lead to federal climate policy. As a global problem reflecting the dangerously elevated concentrations of greenhouse gases in the atmosphere, climate change is most appropriately addressed at the international level through a multilateral agreement and then, failing that, via domestic legislation. Under the former, domestic policy simply fulfills the internationally negotiated commitments. Under the latter, the costs, pace, and plan for a less fossil fuel–intensive energy future is fully subject to debate by a representative domestic decision-making body.

What is unusual about climate change policy in the United States, however, is that the pathway being followed is neither of these but instead one that is arguably the least reflective of the most critical aspects of climate change as a global environmental tragedy of the commons. Federal climate change policy is currently being established through executive branch rulemakings under the 1970 environmental statute, the Clean Air Act, as a result of court mandates rendered in multiple lawsuits filed by states and environmental organizations.

This chapter explores how and why climate change policy making has followed this particular path, with special attention to the role of the courts in the process. The chapter thus chronicles climate change policy making, starting with the delinking of domestic and international climate policy that occurred in the late 1990s with the United States' refusal to ratify the Kyoto Protocol. This delinking established a fixture of domestic federal climate policy making: US legislation would have to sink or swim based upon domestic popularity, unaided by the argument that domestic controls on greenhouse gases are necessary to fulfill international treaty obligations. Despite many attempts and near misses, supporters of climate legislation have been unable to forge the necessary consensus for federal legislation. Instead, the gap left by the lack of congressional action has been filled by rules and regulations issued by the executive branch, specifically the US Environmental Protection Agency, in response to court orders and pursuant to the authority of preexisting environmental statutes. In an interesting twist to the usual scenario in which plaintiff environmental groups dominate litigation to compel federal agencies to promulgate more protective environmental rules, the litigation prompting these executive agency climate change regulatory actions has been spearheaded by states attorneys general.

The chapter will demonstrate that the process of developing federal policy in this manner has resulted in greenhouse gas regulation in fairly quick order, but the regulation avoids, for good or for ill, many of the substantive questions of greenhouse gas policy design, to wit: What regulatory approach (market based, or direct controls upon specific sources?), what targets (only large emitters? land use activities?), and what deadlines (sweeping change now, or gradual change over time, and if the latter, over what time period?) are most appropriate and effective? Current policy also illustrates what scholars have referred to as "bottom up federalism":[1] It is an example of how state and local government actors are filling the climate change policy void created by the lack of federal action.

Climate Change Policy Pathways

Internationally Negotiated Climate Commitments

Climate change is a global environmental problem requiring an international response. The buildup of greenhouse gases results from a global tragedy of the commons: an increase in the concentration of heat-trapping greenhouse gases in the atmosphere as a result of fossil fuel burning and changes in land use practices occurring throughout the world.[2] As with any common pool problem, in the absence of a cooperative solution, economic theory would predict that rational decision makers would fail to take unilateral action, since free riding on the efforts of others would be the preferred solution.[3] Thus commentators, and in particular economists, generally select an international agreement as the preferred policy-making solution to climate change.[4]

Indeed, this has been the pattern with respect to the United States' response to international environmental threats. With respect to such problems, the usual policy-making trajectory is that the United States first participates in negotiating a bilateral or multilateral agreement and then later enacts domestic legislation to implement its international commitments. Such has been the case with respect to efforts over the course of the 20th century to protect migratory birds, prevent the extinction of endangered species, and slow the deterioration of the ozone layer. The United States and Great Britain (on behalf of Canada) negotiated a treaty to protect bird species that migrated across the United States–Canada border.[5] The United States subsequently enacted domestic legislation, the Migratory Bird Treaty Act, to implement the treaty by making it unlawful to pursue, hunt, take, capture, kill, or attempt to take any migratory bird, bird part, or bird egg.[6] The Endangered Species Act[7] implements the Convention on International Trade in Endangered Species of Wild Fauna and Flora.[8] Under the Montreal Protocol, developed nations agreed to the phaseout of ozone layer–depleting chlorofluorocarbons.[9] The Clean Air Act was subsequently amended in 1990 to implement the phase-out within the United States.[10]

With respect to climate change, however, the United States has not pursued the international treaty followed by domestic legislation. As Yael Wolinsky-Nahmias discusses in the Introduction to this book, in 1992, the US Senate ratified the United Nations Framework Convention on Climate Change (UNFCCC). While this action raised the expectation that the United States would continue to participate in international negotiations, ultimately developing a domestic greenhouse gas mitigation program pursuant to an international treaty, the UNFCCC actually constituted the zenith of the United States' commitment to an international climate change agreement. Five years later, the US Senate would pass, by unanimous vote, the Byrd-Hagel Resolution, prohibiting the United States from becoming party to any international agreement requiring specific reductions in domestic greenhouse gas emissions without also requiring reductions from developing country parties, and from becoming party to any agreement that would result in "serious harm to the economy of the United States."[11] While the Clinton administration was sensitive to the interests of environmentalists who were calling for climate change regulation, the influence of natural resource extraction industries in Congress presented a political challenge: 26 out of 50 states have significant coal extraction.[12] The economic impacts of greenhouse gas regulation on these industries, and thus these states, could be dramatic. Given this political context, it is perhaps not surprising that, although President Bill Clinton signed the 1997 Kyoto Protocol, he never went on to submit the Kyoto Protocol to the Senate for ratification.

While the Obama administration has made nonbinding commitments to reduce greenhouse gas emissions in international forums,[13] it has rejected, as politically infeasible, an approach requiring binding commitments by the United States or any other developed nation.[14] For the foreseeable future, therefore, the ongoing international climate law-making process will continue to be but a weak driver of domestic policy on climate change.

Unilateral Domestic Climate Policy

It is often assumed that it would be economically irrational for a party to unilaterally reduce the intensity of its beneficial use of a commons resource out of a desire to prevent the degradation of the commons. Although a commons user will benefit individually from a healthier commons, the prospect of free riding on the reductions of other commons users is assumed to stop a party from acting against his or her self-interest in such a manner.[15] To the contrary, it is economically rational for a party whose contribution to commons' degradation is sufficiently great— that is, great enough that some amount of commons' degradation is attributable to it alone—to undertake unilateral action to reduce its footprint. The reason for this is that the benefits that accrue to such large contributors from a healthy commons resource more than offset the harm that accrues from a reduction in the intensity of their use.

This dynamic is arguably the case with respect to the United States and other major national contributors to the buildup of global greenhouse gas concentrations responsible for the degradation of our climate system "commons." The United States is responsible for approximately 15 percent of global carbon dioxide emissions from fossil fuel combustion, an amount exceeded only by China at 29 percent.[16] These emissions are far from costless; climate change is estimated to significantly lower the United States' gross domestic product (GDP). Interestingly, economic studies suggest that the United States has more to gain than lose from unilateral action to reduce its greenhouse gas emissions. The cost of emissions reductions is smaller than the amount by which climate change negatively affects the nation's GDP.[17]

Although the current buildup of greenhouse gases in the atmosphere represents the aggregation of emissions from all countries in the world, some nations have contributed more than others historically. Figure 2.1 shows the cumulative percentage contribution of GHG emissions by various nations, reflecting all GHG emissions known between 1850 and 2010. Note that no single lifetime for carbon dioxide can be defined because of the different rates of uptake by the

FIGURE 2.1   Contributions to Cumulative Greenhouse Gases Emissions (percentage), 1850–2010

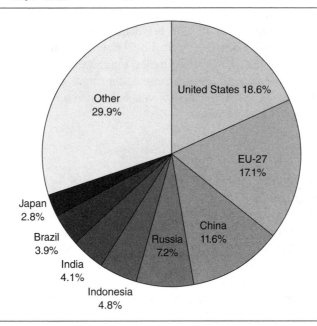

Data Source: Michael D. J. den Elzen, Jos G. J. Olivier, Niklas Höhne, and Greet Janssens-Maenhout. "Countries' Contributions to Climate Change: Effect of Accounting for All Greenhouse Gases, Recent Trends, Basic Needs and Technological Progress," *Climactic Change* 121 (2013): 402.

different physical removal processes, such as the absorption of carbon dioxide by the oceans.

For the United States, the policy-making pathway of unilateral mitigation has been anything but smooth, however. For most of the last 25 years, federal lawmakers have preferred to invest in climate-related research, defeating, several times on narrow grounds, any regulatory regime imposing mandatory cuts on domestic greenhouse gas emissions. Concerns about the implications for such a regime on the economy have led Congress to consider mainly cap-and-trade bills that would enable industry to comply through the purchase of emission allowances from other regulated parties or through the funding of greenhouse gas reduction projects in developing countries. Such flexible regulatory approaches nevertheless fell victim to concerns over negatively impacting the economy and fears over the competitive disadvantage of doing so in the absence of the adoption of similar curbs in the rapidly expanding economies of developing nations, particularly China and India.

When scientific awareness of the association between human activities and changes to the Earth's climate emerged in the 1970s, Congress established research programs to better understand the mechanisms of climate change. As early as 1978, Congress created a national climate program, the purpose of which was to "assist the Nation and the world to understand and respond to natural and man-induced climate processes and their implications."[18] Congress's involvement in climate change research began in earnest in 1990 with legislation establishing the US Global Change Research Program.[19] The act authorizing this program directed the president to establish a 10-year research program to "improve understanding of global change"[20] and provided for periodic scientific assessment that analyzes "current trends in global change, both human-induced and natural, and projects major trends for the subsequent 25 to 100 years."[21] The Act furthermore established the Office of Global Change Research Information, the purpose of which was to disseminate the scientific research useful in "preventing, mitigating, or adapting to the effects of global change."[22] To that end, the office was specifically charged with distributing information for reducing the amount of greenhouse gases released to the atmosphere through conservation, energy efficiency, promotion of renewable energy sources, and conservation of forest resources.[23]

The policy landscape of the early 2000s sharply contrasted with the federal government's support for climate change research during the 1990s. The first decade of the millennium was marked by the open hostility of the Bush administration to the Kyoto Protocol and the administration's renunciation of any plans to establish carbon dioxide emissions reductions for US power plants. Congressional energy bills packed with subsidies to fossil fuel interests dominated the domestic policy landscape in the early 2000s. The 2005 energy bill signed into law contained a $2.6 billion package of oil and gas exploration incentives and $3 billion of subsidies for coal extraction.[24] The bill included new incentives for oil drilling in the Gulf of Mexico, new subsidies to develop ethanol and "clean coal" technology, and an exemption for fracking from the underground injection control program of the Safe Drinking Water Act. (With the more recent rise in the use of fracking to extract natural gas, many now believe this exemption is endangering water supplies in many areas of the United States.) Despite a new democratic majority in Congress by 2007, the 2007 energy bill failed to pass a proposal for $32.2 billion of tax increases for the domestic oil and gas industry that were to fund $32.2 billion of tax cuts for alternative fuels and energy conservation.

Data compiled by the League of Conservation Voters highlights a polarization between politicians with respect to environmental issues that exceeds divisions among voters. Voting patterns since the 1970s show that citizen support for increased environmental spending does not diverge along party lines nearly as much as it diverges from the support provided by legislators (see Figure 2.2). Such high polarization among political leaders helps explain the lack of federal climate policy.

In the negative climate policy environment of the Bush administration, the Climate Stewardship Act of 2003, also known as the McCain–Lieberman bill after

FIGURE 2.2   Congressional Proenvironment Scores and Citizen Support for Increased
Environmental Spending 1970s–2000s

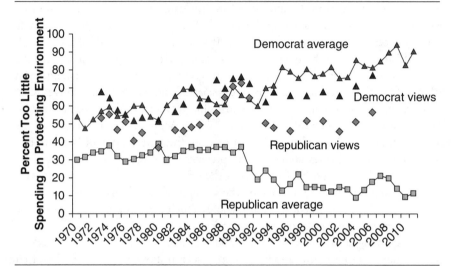

Source: Theda Skocpol, *Naming the Problem: What It Will Take to Counter Extremism and Engage Americans in the Fight Against Global Warming,* Prepared for the Symposium on The Politics of America's Fight Against Global Warming, Harvard University, Cambridge, MA, February 14, 2013, 58 (with permission).

its bipartisan cosponsors Senators John McCain (R-AZ) and Joseph Lieberman (D-CT), was a notable exception.[25] Not surprisingly, however, it failed to pass Congress. Opposed by both the chair of the Senate Environment and Public Works Committee and by President Bush, the bill failed to garner a Senate majority despite being introduced three times. The Act would have covered a sizable fraction of domestic greenhouse gas emissions, capping emissions and allowing regulated parties to comply with their emission reduction obligations by purchasing emission allowances or demonstrating, through activities registered with the National Greenhouse Gas Database, a net increase in carbon sequestration.[26] The successor to the 2003 Act, the 2007 Climate Security Act, would have capped emissions at 2005 levels and thereafter gradually reduced emissions to 63 percent below 2005 levels by 2050. This Act also failed to gain passage in Congress.

The election of President Barack Obama prompted a renewed effort in Congress to enact climate legislation. Two bipartisan efforts eventually failed, however, despite making impressive gains in obtaining political support. Another bill crafted by McCain and Lieberman in 2009 looked promising until McCain, facing a tough reelection bid, backed away from the bill.[27] A second effort led by Lieberman, John

Kerry (D-MA), and Lindsey Graham (R-SC), contained a "grand bargain"—a cap on emissions levels, increased support for nuclear power, and sweeping expansions in offshore oil drilling. Nevertheless, while the House passed the compromise-laden American Clean Energy and Security Act of 2009, similar legislation failed to pass the Senate. It may never be possible to untangle all of the competing theories for this failure, but it is important to consider that 2009 marked the start of the nation's financial crisis, a sharp rise in unemployment, and a Republican wing of Congress facing increased pressure from Tea Party activists hostile to cap-and-trade and indeed any form of climate legislation.[28]

While the financial crisis may have contributed to the failure of climate legislation, it did spur major investments in renewable energy and efficiency programs. In contrast to the tax breaks for the fossil fuel industry found in the energy bills of 2005 and 2007, the American Recovery and Reinvestment Act of 2009 contained $27.2 billion in grants, loan guarantees, and other incentives and supports for renewable energy research and investment and energy efficiency. Programs included $6 billions for renewable energy and electric transmission technology loan guarantees, $5 billion for weatherizing low-income homes, $3.4 billion for carbon capture and low-emission coal research, and $3.2 billion in energy efficiency and conservation block grants to states, tribal authorities, and local governments.

During President Obama's second term of office, advocates are again clamoring for Congress to act on climate legislation. The president has asked for congressional action, threatening, "If Congress won't act soon to protect future generations, I will."[29] Because the Obama administration in fact has been addressing climate change through executive actions, this threat has real credibility. In fact, such executive actions play a central role in the administration's 2013 energy plan, which obligates the United States to reducing greenhouse gases by 17 percent below 2005 levels by 2020.[30] This ambitious plan commits the executive to promulgating carbon pollution standards for new and existing power plants between 2013 and 2016, adopting new efficiency standards for appliances and in federal buildings, encouraging investment in climate resiliency, and providing greater engagement on climate mitigation internationally.

Judicial Climate Policy Pathways

A final policy pathway for addressing climate change is through court decisions. The many types of lawsuits that can be filed in an attempt to achieve progress in climate mitigation are shown in Figure 2.3.

The climate cases being filed in the courts reflect a great diversity of claims and targets. The largest number of cases, representing roughly one-quarter of the 527 cases filed as of October 2012, concern challenges to federal action, primarily the federal government's failure to regulate greenhouse gas emissions. The second-largest

FIGURE 2.3   Types of Climate Cases Filed 1989–2012

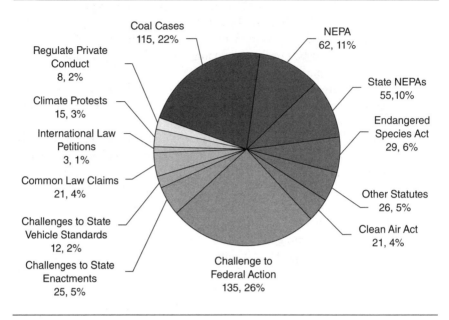

Source: Michael B. Gerrard, "Climate Change Litigation in the United States", www.climatecasechart .com (with permission).

category comprises cases filed against coal-fired power plants, and the third is actions under the National Environmental Policy Act, the federal environmental impact assessment statute, which are roughly equal in number to those filed under state NEPA laws.

This section will highlight three basic types of cases. In the first, owners of private property or government entities, such as states, in their role as stewards over property, can file actions directly against large sources of greenhouse gases. The plaintiffs in such cases often claim that these sources are harming their property and seek a court order to have the emitting source reduce the size of its emissions or pay damages for the harm it has caused their property. In the second, parties may file suit contending that a government agency is legally obligated to regulate the emission of greenhouse gases from a whole class of sources. In the third, parties can file suit against government agencies claiming that these agencies must either deny permits to sources due to their excessive emission of greenhouse gases or due to their failure to adequately analyze the environmental impacts of their greenhouse gas emissions. Although the first type—those claiming property damage—have generated the most publicity, it has been the second and third types that have been the most successful and have led to what is now the backbone of the federal response to climate change.

Public Nuisance Litigation

Perhaps the boldest use of the courts to attempt climate mitigation has been the common-law nuisance claims filed by states and environmental groups directly against large sources of greenhouse gas emissions. In these actions, the plaintiffs sought court-ordered reductions in the defendants' emissions or damages attributable to the source's emissions. Each of the lawsuits was premised upon a common-law liability theory, most significantly, that the defendant's emissions constituted a public nuisance—an unreasonable interference with a right general to the common public.[31]

While the amount of media coverage might lead one to believe dozens of nuisance cases were filed, this legal tactic was the basis of just four lawsuits. In the first, California filed suit against General Motors and five other car manufacturers, alleging that the emissions of greenhouse gases from the automobiles contributed to the public nuisance of climate change, characterized by an enhanced wildfire danger as well as the early melting of the snowpack that is so critical to supplying the state's water needs.[32] California requested both injunctive relief, in the form of a court order to the manufacturers to reduce their emissions, and monetary damages for harms attributable to the emissions. In the second, California, and several mostly northeastern states, filed suit against six of the largest electric utilities in the United States, alleging that the companies' greenhouse gas emissions constituted a public nuisance.[33] In the third, families living on the Mississippi coast that experienced losses attributable to Hurricane Katrina in 2005 filed a public nuisance action against oil and gas, coal, and chemical production companies, alleging that the damages they sustained from the hurricane had been aggravated by the defendants' greenhouse gas emissions. In the fourth action, a Native Alaskan village filed suit against fossil fuel production companies, alleging that the companies' emissions had accentuated climate change, which was in turn responsible for sea level rise and other climatic changes that would interfere with the villagers' way of life and require them to relocate their settlement.[34] The village requested the court grant them the costs of relocation, an expense estimated at $400 million.

Many of the nuisance actions just mentioned were dismissed at the trial court level on the ground that the action presented a "political question" better left for resolution by the political branches of government as opposed to the courts. In the action filed by states against electric utilities, for instance, the court held that the plaintiff's request for relief—an order requiring the utilities to reduce their emissions—required a balancing of economic, environmental, foreign policy, and national security interests. Therefore, the case could not be decided in the absence of an initial policy decision by the elected branches of government.[35] Among the questions the court believed must first be answered by a political branch were the following: "Given the numerous contributors of greenhouse gases, should the societal costs of reducing such emissions be borne by just a segment of the electricity-generating industry and their industrial and other consumers?" and "What are the implications for the nation's energy independence and, by extension, its national security?"[36]

Not all courts agreed that the climate nuisance cases were incapable of resolution in the absence of prior political action. In particular, the appeals court in *Connecticut v. American Electric Power Co.,*[37] the states' utility public nuisance case, stated that the need for an initial policy decision from a political branch was belied by the courts' successful adjudication of complex common-law nuisance cases for over a century, well before legislatures began addressing environmental matters. The court in *American Electric Power* stated that there was no reason to be afraid of courts deciding climate nuisance cases: "[A] decision by a single federal court concerning a common law of nuisance cause of action, brought by domestic plaintiffs against domestic companies for domestic conduct, does not establish a *national* or *international* emissions policy."[38]

Many legal commentators, on the other hand, were critical of the use of the public nuisance strategy to address climate change on the grounds that the cause of action was ill suited to the problem of climate change. For one, common-law liability requires that the plaintiff be able to prove that his or her harm was caused by the defendant's action. This is a near-impossibility with respect to greenhouse gas emitters, since the contribution of a single source of greenhouse gas emissions to climate change eludes scientific approximation. According to one commentator, "Private and public nuisance cases have not been used to manage the load over time on common natural resources, and for good reason: courts are not well-equipped to generate and implement long-term resource management plans."[39]

Using the doctrine of "displacement," the Supreme Court ultimately dealt the federal public nuisance cause of action a death blow. Under this doctrine, the court's authority to decide an issue is precluded where Congress has already provided a rule of decision on the matter. In *American Electric Power Co. v. Connecticut* the Court found determinative the holding of *Massachusetts v. EPA,* decided four years previously: that greenhouse gases are subject to regulation under the Clean Air Act. Despite the fact that EPA had yet to actually use this authority to regulate greenhouse gas emissions, the Court stated that the "critical point is that Congress delegated to EPA the decision whether and how to regulate carbon-dioxide emissions from power plants; the delegation is what displaces federal common law."[40] As a result of this decision, any lawsuit seeking court-ordered cutbacks in emission of greenhouse gases otherwise subject to regulation under the Clean Air Act is similarly displaced. Evidence of this is provided by the action of the Alaskan Native village, also under the federal public nuisance doctrine. The case was dismissed by the appeals court on the basis that it was preempted by the Clean Air Act.[41] All that remains of public nuisance is the possibility that a party might file an action under state public nuisance doctrine, a scenario left open by the Supreme Court but one also fraught with legal complexities.

At the time they were ongoing, these lawsuits grabbed headlines with their vivid portrayals of the harms associated with climate change. There were also indications that elements of the fossil fuel industry considered the lawsuits to represent a significant threat to its interests and thus perhaps that these lawsuits could have been instrumental in obtaining important concessions from industry in national legislation.[42] Nevertheless, the courts proved uncomfortable with their role in these lawsuits and in general adopted legal rulings that preclude such lawsuits under federal law, ceding primary authority to regulate climate change to the legislative and executive branches of government. It is unclear what role, if any, these lawsuits have had in the overall national climate policy-making debates.

Court-Mandated Executive Branch Regulatory Actions

At the same time that Congress was debating climate legislation, environmental groups and states within the United States were contemplating court actions to compel the regulation of greenhouse gas emissions under existing laws. Ultimately, it has been this effort, which has culminated in executive action to regulate greenhouse gas emissions under the federal Clean Air Act, which now constitutes the cornerstone of the federal response to climate change.

A watershed decision by the US Supreme Court, ultimately titled *Massachusetts v. Environmental Protection Agency*, led the way. This suit had its origin in a rulemaking petition filed by 19 environmental organizations requesting that the EPA regulate the emission of greenhouse gases from new motor vehicles under the Clean Air Act.[43] The act states that EPA administrator "shall" prescribe standards applicable to the emission of "any air pollutant" from new motor vehicles "which in his judgment cause, or contribute to, air pollution which may reasonably be anticipated to endanger public health or welfare."[44] Environmental groups quickly determined that, if EPA were to determine that greenhouse gases are "air pollutants," EPA would be required to establish emission standards for cars and trucks. The US transportation sector generally accounts for over a quarter of the United States' greenhouse gas emissions and approximately 7 percent worldwide.[45] As a result, the amount of greenhouse gases potentially reduced through this provision was significant.

A threshold issue posed by the petition and later resolved by the Supreme Court was whether emissions of greenhouse gases were within the scope of EPA's power to regulate under the Clean Air Act. EPA had previously maintained that they were. In 1998, Jonathan Z. Cannon, the general counsel of EPA at the time, submitted a memorandum to then–EPA Administrator Carol Browner, stating that "CO_2 emissions are within the scope of EPA's authority to regulate," though the agency had never actually exercised its authority to do so.[46] Mr. Cannon's successor, Gary Guzy, repeated this statutory interpretation in testimony before a congressional committee, stating, "Given the clarity of the statutory provisions defining 'air pollutant' and providing authority to regulate air pollutants, there is no statutory ambiguity" concerning whether EPA can regulate carbon dioxide under the Clean Air Act.[47]

Despite these previous statements by high-level EPA officials, when the International Center for Technology Assessment, a technology information advocacy group, discouraged by the lack of action by Congress, attempted to use the Clean Air Act to compel regulation of greenhouse gases, they were met with significant resistance from the administration of President George W. Bush. The EPA denied the group's rule-making petition, stating that the generality of the Act's reference to "any air pollutant" and Congress's amendment of the Act to address the deterioration of the ozone layer demonstrated that Congress did not intend for the Act to apply to a global problem such as climate change, but rather only to more local or state-level problems, such as urban smog or regional haze.[48] The agency also argued that regulation of greenhouse gases would disadvantage the United States vis-à-vis other nations in international climate negotiations, since the United States would not be able to use domestic regulation as a bargaining chip, and that "scientific uncertainty about the causal effects of greenhouse gases on the future climate of the earth" warranted delaying regulation until a greater consensus on the scientific evidence was reached.[49]

Massachusetts and other states appealed the EPA's denial and, upon receiving a negative decision there, sought review in the US Supreme Court. The closely divided Supreme Court decision was a watershed opinion on at least two levels. First, the Court ruled in favor of Massachusetts on the hotly contested side issue of whether the state had "standing" to sue, a constitutional prerequisite for any litigant to bring suit in federal court. Here standing required that Massachusetts establish the requisite "injury in fact;" that it could trace its injury to the EPA's failure to regulate greenhouse gases from new mobile sources, and that the Court's actions mandating the agency regulate such emissions, assuming the agency subsequently found vehicle emissions "may reasonably be anticipated to endanger public health or welfare," would redress Massachusetts's injury.

While the harm inherent in the loss of coastline due to sea level rise was easy for the Court to swallow, the knottier issue was whether regulation of just 6 percent of world emissions caused by new motor vehicles would redress this injury, given the likelihood that countries such as China and India are only increasing their emissions, and hence sea levels are virtually certain to rise regardless of whether EPA regulates new vehicle emissions or not. The Court cited a number of considerations in holding that states, as sovereigns, were entitled to a special status under the doctrine that litigants must be injured to invoke the power of the courts. Accordingly, the Court held that Massachusetts was entitled to make out the requisite redressability demonstration by showing that, in the absence of court action, Massachusetts was powerless to vindicate its sovereign interests in the protection of its territory, and that a reduction in domestic greenhouse gas emissions would slow the pace of global warming regardless of what other countries did.[50] With respect to the latter consideration, the Court held that it was important to recognize that federal agencies seldom resolve massive global problems "in one fell regulatory swoop" but instead "whittle away at them over time, refining their preferred approach as circumstances change and as they develop a more nuanced understanding of how best to proceed."[51]

While the standing holding established an important precedent with respect to state-led climate litigation, especially against the federal government, the Court's holding on the merits paved the way for the EPA's subsequent regulation of greenhouse gases under the Clean Air Act more generally. Here Massachusetts's statutory argument found favor with the justices who have tended to favor a "plain meaning" approach to statutory interpretation. The Court held that, because the Clean Air Act's "sweeping definition" of an air pollutant on its face embraces "all airborne compounds of whatever stripe," greenhouse gases were "without doubt" air pollutants under the Act.[52]

The Court's holding that greenhouse gases are "air pollutants" under the Clean Air Act has had the effect of enabling the EPA—albeit after it made other necessary findings—to institute greenhouse gas mitigation regulatory programs under a variety of provisions of the Act. To actually regulate carbon dioxide emissions from motor vehicles, it was necessary that the EPA first find that emissions of such gases endanger public health or welfare and that the combined emissions of such gases from new motor vehicles "cause or contribute" to air pollution, which may reasonably threaten human health and welfare. Despite initial resistance by the White House to an affirmative finding of endangerment, the EPA moved forward with both findings, eventually issuing them in final form in December 2009.[53] In making its determination, the EPA relied upon research by the US Global Change Research Program, the Intergovernmental Panel on Climate Change, and the National Research Council of the US National Academy of Sciences to substantiate its claim that climate is changing due to the buildup of heat-trapping greenhouse gases attributable primarily to human activities, and that these climate changes will endanger both public health and public welfare.

While the EPA's endangerment finding triggered a host of legal challenges, it serves as the necessary predicate to further federal and, interestingly enough, state regulation under the Clean Air Act. As a result of this finding, the EPA, in collaboration with the National Highway Traffic Administration, issued the greenhouse gas emission standards for motor vehicles originally sought by the environmental organizations in their

1999 rule-making petition.[54] This has been followed by other regulatory actions under the Clean Air Act. Likely the most significant is EPA's standards for new electricity-generating plants.[55] The rule caps such plants' carbon dioxide emissions at 1,000 pounds per megawatt hour, a standard easily met by natural gas plants, but too low to be met by coal-fired plants in the absence of carbon sequestration technology which has yet to become commercially viable. As a result, this EPA rule essentially eliminates the prospect of new coal-fired utility plants. Finally, EPA has issued emission thresholds to define when sources of greenhouse gas emissions must obtain a permit under the Act's "prevention of significant deterioration" program, a program requiring plants install the best available control technology.[56] These emission thresholds make that program workable by ensuring that the permit program applies only to larger sources of greenhouse gases.

States and environmental groups have not limited the tactic of using existing environmental statutes to address climate change to the Clean Air Act. Nonprofit environmental organizations, active in litigation under the Clean Air Act, have also been pushing for government regulation of greenhouse gas emissions under other statutes, including the Endangered Species Act, the Clean Water Act, and the Energy Policy Act.[57] Litigation under these other statutes consists primarily of challenges to particular government decisions, but others seek across-the-board regulation to reduce greenhouse gases or address the impacts of climate change.

Examples of the latter can be found with respect to wildlife and water protection. For example, the Center for Biological Diversity, an environmental group active in urging the government to protect the environment from climate change, sued the EPA for failing to require Washington State to list its marine waters as impaired waters under the Clean Water Act due to the ocean's absorption of greater quantities of greenhouse gases.[58] The EPA settled the lawsuit by agreeing to issue guidance assisting states in identifying acidic coastal waters. Even more important, considering the scope of the ruling, in 2008 the US Fish and Wildlife Service listed the polar bear as a "threatened" species under the Endangered Species Act, based largely upon the threats to its sea-ice habitat due to warming global temperatures. Environmental groups have filed dozens of petitions asking the service to list species as threatened or endangered due to evidence that the species' habitat is dwindling as a result of climate change impacts.

Environmental Assessment

In a third type of climate lawsuit, parties seek to enforce the requirements of the National Environmental Policy Act (NEPA) with respect to projects that have a potential to increase climate change. Under NEPA, federal agencies are obligated to consider the environmental impacts of any proposed action upon the quality of the human environment. Where that impact is considered to be significant, the agency is required to prepare an Environmental Impact Statement that assesses the impacts of, alternatives to, and mitigation of the proposed action. NEPA is considered to be a procedural, as opposed to a substantive, environmental statute. The law requires federal agencies to analyze the environmental impacts of their decisions prior to taking final action, but it does not prohibit the agency from following through on actions, even those found to produce harmful environmental impacts. The point of the statute is to inform officials of the impacts of the proposed action and of alternatives to the proposed action, so they may use their discretion to adopt a less harmful action and to inform the public of the choices before the agency. Thus the Act is read to require agencies "stop, look, and listen" prior to acting in a way that will impact the environment, but it does not dictate the agency's ultimate actions.

The failure of an agency to properly consider climate change as an environmental impact in Environmental Impact Statements has been the motivation for the majority of cases brought against the federal government. The actions at issue in these cases range from the National Highway Traffic Safety Administration's fuel economy standards for light trucks[59] to presidential right-of-way permits for construction of electricity transmission lines transporting power from Mexican power plants to southern California[60] and the proposed construction of rail lines to carry coal out of Wyoming's Powder River Basin.[61]

The NEPA case law contains some notable rulings that get at some of the difficulties of accounting for the environmental impacts of government actions that facilitate the burning of fossil fuels. In one appellate court decision, the court held that the agency must address "reasonably foreseeable" impacts, even if only the nature of the impact is known and not its extent.[62] Thus an agency must examine the environmental impacts flowing from an increase in coal consumption—a reasonably foreseeable result of the agency's construction of a rail line to carry low-sulfur coal from the Powder River Basin.[63] A second issue is how to account for the scientific fact that climate change results from aggregate global levels of greenhouse gases to which each individual agency action contributes just a small fraction of such gases. The court held that the duty to analyze the cumulative environmental effects of agency actions under NEPA meant that agencies must analyze the impacts of greenhouse gas emissions resulting from the action at issue plus other related rule makings and "other past, present, and reasonably foreseeable future actions, regardless of what agency or person undertakes such other actions."[64]

Implications of the Current Status of Federal Government Regulation and the Use of the Courts

The lack of federal legislation specifically addressing climate change and the prominent role played by the courts in triggering regulation have had profound implications for federal climate mitigation policy. The following specifically identifies the most influential aspects of this predicament and what each has meant for the current status of climate change regulation.

Reliance upon existing legal authorities—Obviously the most important influence on the current status of climate change regulation is the absence of a specific congressional act addressing climate change and the consequent reliance upon existing legal authorities to address climate change. The result of the use of existing authorities is the fairly quick implementation of rules reducing greenhouse gas emissions, but with it a somewhat erratic, ad hoc scheme of regulation.

While the court-mandated greenhouse gas emission rules discussed in this chapter are still being developed by government agencies, they are on a far faster timetable than congressional legislation that has still failed to materialize. This manner of rule development also avoids the inevitable delays attending the give-and-take of the legislative process. For example, the 2009 Waxman–Markey climate bill passed by the House of Representatives mandated a reduction in greenhouse gas emissions of 17 percent below 2005 levels by 2020 and 83 percent below 2005 levels by 2050. The first target imposed a real mandate to begin cutting emissions almost immediately, but the "83 percent" "by 2050" reeks of someday, somehow intentions, tempting regulators to put off the medicine until the half-century mark was all but looming on the horizon. Thus one benefit of the "climate change policy by litigation rather than legislation" is that the emissions standards being put into effect have shorter-term, more certain deadlines for action.

On the downside, however, are other consequences of the litigation strategy; among them the resources spent by the EPA and others in litigating the cases. The use of existing legal authorities to address climate change that were not drafted with climate change specifically in mind sets up the predictable pattern of litigation illustrated by the history of legal actions on this topic. Given that most environmental statutes are implemented by an administrative agency—generally the EPA or the Department of the Interior—within or controlled by the president, the parties driving the litigation will depend upon whether the administration is favorably or unfavorably inclined to apply the existing legal authorities to a problem unanticipated by the Congress that enacted the legislation being implemented. Thus in the case of the use of the Clean Air Act to address climate change, the administration in office from the late 1990s through the mid-2000s—that of George W. Bush—was hostile to the idea of mandatory regulation, preferring voluntary greenhouse gas reductions instead. As a result, states and environmental organizations were the initiators of litigation seeking to compel the government to apply its authority under the act to regulate greenhouse gases. Such litigation is less apt to be successful, given the difficulties of establishing constitutional standing to sue and compelling a government agency to act. *Massachusetts v. EPA* was thus an unusual case in which certain factors enabled the plaintiffs to overcome these barriers.

A different pattern of litigation emerges where the administration is favorably inclined to use existing legal authorities to regulate. This is the circumstance that has been prevailing since the Obama administration took office in 2008, in which the EPA has actively sought to control greenhouse gases through all of the authorities available under the Clean Air Act. Here the initiators of litigation are the targets of EPA regulation (and their supporters), who generally contend that the EPA either lacks the authority to regulate or that the EPA's regulations are arbitrary and capricious. Thus the EPA's 2009 finding that emissions of greenhouse gases endanger public health or welfare, which serves as a prerequisite to regulation of emissions of a pollutant under multiple provisions of the Act, has attracted multiple legal challenges. These challenges were filed by industries and commercial interests potentially subject to future climate change regulation as well as by states sympathetic to those interests, which are mainly related to the fossil fuel industry.[65]

Another cost resulting from the use of existing authorities to regulate greenhouse gas emissions is the resources spent by agencies adjusting existing legal authorities to the problem it is trying to fix. Again, the EPA's experience using the Clean Air Act to address climate change is a case in point. Under the Act, stationary sources emitting more than threshold amounts of a pollutant regulated under the Act must meet stringent emission standards in order to receive an operating permit. While this provides a potent authority for reducing greenhouse gas emissions from coal-fired power plants and other stationary sources, the permitting thresholds are established by statute at levels that, while appropriate for triggering technological controls of conventional pollutants, are far too low to serve as trigger levels for controlling greenhouse gas emissions. Thus, in order to avoid sweeping into the scope of the mandatory permitting programs literally thousands of facilities, including, allegedly, farms, ranches, and doughnut shops, the EPA issued a rule raising the statutory thresholds by regulation.[66] Despite having the effect of easing a statutory burden, because of the EPA's uncertain legal authority to change, via regulation, explicit statutory thresholds, its "tailoring rule" attracted multiple legal challenges. While the rule was ultimately upheld,[67] it was only after the EPA spent significant resources to defend the rule.

Finally, and perhaps most important, an existing regulatory framework closes off certain regulatory design options. Some of the most prevalent designs for national greenhouse gas legislation—a national cap-and-trade scheme—are not possible, or at least not clearly possible, under the Clean Air Act. While some herald direct emissions reductions mandated by the existing Clean Air Act scheme, others see it as inefficient. The benefit of a cap-and-trade regime is the opportunity to obtain emissions reductions from those facilities for whom reductions are cheapest. The cap-and-trade regimes considered by Congress also contemplated giving credit for emissions reductions to sources that paid for reductions occurring in developing countries. This is also not possible under the ad hoc regulatory regime resulting from the use of current statutory authorities.

Conclusion

The history of the development of federal climate policy has followed an unexpected path. Contrary to the predictions of many, Congress did not step up during the late 1980s, or indeed at any point since, to enact comprehensive climate legislation despite exhaustive consideration of many competing legislative proposals and several close votes. In the absence of federal legislation, the executive branch, goaded into action by lawsuits filed by public interest groups and state governments, has responded to the calls for regulation through piecemeal regulations under existing statutory authorities, most prominently the Clean Air Act, but also the Clean Water Act, the Endangered Species Act, and the National Environmental Policy Act. Climate policy making that is most accurately termed "regulation via existing statutes" presents a mix of advantages and disadvantages. To some, avoiding the interest group–laden politics of climate change in Congress and the resulting legislative deals is a clear benefit of the current scheme. As a result of the current scheme driven by court deadlines and existing statutory deadlines, there is also a good argument that regulation has come sooner, though clearly in a less comprehensive format. On the other hand, the piecemeal nature of the regulatory scheme that has resulted and the difficulties inherent in applying, to climate change, statutory provisions designed for other environmental problems, has short-circuited a more rational and comprehensive regulatory response to climate change.

Suggested Readings

Engel, Kirsten H. "Courts and Climate Policy: Now and in the Future." In *Greenhouse Governance Addressing Climate Change in America,* edited by Barry G. Rabe, 10–43. Washington, DC: Brookings Institution Press, 2010.

Freeman, Jody, and Andrew Guzman. "Climate Change and US Interests." *Columbia Law Review* 109 (2009): 1531–1601.

Markell, David, and J. B. Ruhl. "An Empirical Assessment of Climate Change in the Courts: A New Jurisprudence or Business as Usual?" *Florida Law Review* 64 (2012): 15–86.

References

Alexander, Kristina. *The Endangered Species Act as (ESA) as Implementing Legislation for International Treaties.* CRS Report to Congress (2012). https://www.hsdl.org/?view&did=707792.

American Electric Power Company v. Connecticut, 131 S. Ct. 2527 (2011).

Engel, Kirsten H. "Courts and Climate Policy: Now and in the Future." In *Greenhouse Governance: Addressing American Climate Change Policy,"* edited by Barry G. Rabe, 229–259. Washington, DC: Brookings Institution Press, 2010.

Fisher, Dana R. "Bringing the Material Back In: Understanding the US Position on Climate Change." *Sociological Forum* 21 (2006): 467, 481.

Freeman, Jody, and Andrew Guzman. "Climate Change and US Interests." *Columbia Law Review* 109 (2009): 1531–1601.

Giddens, Anthony. *The Politics of Climate Change.* Cambridge, MA: Polity, 2009.

Hoffman, Matthew J. *Climate Governance at the Crossroads.* Toronto: University of Toronto Press, 2011.

Lutsey, Nicholas, and Daniel Sperling. "America's Bottom-Up Climate Change Mitigation Policy." *Energy Policy* 36 (2008): 671, 674.

Markell, David, and J. B. Ruhl. "An Empirical Assessment of Climate Change in the Courts: A New Jurisprudence or Business as Usual?" *Florida Law Review* 64 (2012): 15–86.

Massachusetts v. Environmental Protection Agency, 549 U.S. 497, 523 (2007).

Rabe, Barry G., ed. *Greenhouse Governance: Addressing Climate Change in America.* Washington, DC: Brookings Institution Press, 2010.

Stavins, Robert N. *Policy Instruments for Climate Change: How Can National Governments Address a Global Problem?* Discussion Paper 97-11, prepared for the University of Chicago Legal Forum. Washington, DC: Resources for the Future, 1997.

Victor, David G. *Global Warming Gridlock: Creating More Effective Strategies for Protecting the Planet.* Cambridge, UK: Cambridge University Press, 2011.

Notes

1. Nicholas Lutsey and Daniel Sperling, "America's Bottom-Up Climate Change Mitigation Policy," *Energy Policy* 36 (2008): 674.
2. B. Metz, O. R. Davidson, P. R. Bosch, R. Dave, and L. A. Meyer, eds., *Climate Change 2007: Mitigation of Climate Change.* Contribution of Working Group III to the Fourth Assessment Report of the Intergovernmental Panel on Climate Change (Cambridge, UK, and New York: Cambridge University Press, 2007), 2.2.1.
3. See Mancur Olson, *The Logic of Collective Action: Public Goods and the Theory of Groups* (Cambridge, MA, Harvard University Press, 1965).
4. See, e.g., Robert N. Stavins, "Policy Instruments for Climate Change: How Can National Governments Address a Global Problem?" Discussion Paper 97-11, prepared for the University of Chicago Legal Forum. Washington, DC: Resources for the Future (1997). ("Because unilateral action will invariably be highly inefficient, any domestic program requires an effective inter-national agreement, if not a set of international greenhouse policy instruments," p. 323.)
5. Convention for the Protection of Migratory Birds, August 16, 1916, United States–Great Britain (on behalf of Canada), 39 Stat. 1702, T.S. No. 628.
6. 16 U.S.C. §§ 703–711.
7. 16 U.S.C. §1537a.
8. 27 UST 1087; TIAS 8249; 993 UNTS 243. See also Kristina Alexander, *The Endangered Species Act (ESA) as Implementing Legislation for International Treaties,* CRS Report to Congress (2012), https://www.hsdl.org/?view&did=707792.

9. Montreal Protocol on Substances That Deplete the Ozone Layer, 1522 UNTS 3; 26 ILM 1550 (1987).

10. 42 U.S.C. 7401 et seq.

11. See S. Res. 98, 105th Cong. (1997).

12. Dana R. Fisher, "Bringing the Material Back: Understanding the U.S. Position on Climate Change," *Sociological Forum* 21 (2006): 467, 481.

13. In 2009, the United States signed the Copenhagen Accord, a nonbinding climate change agreement that the United Nations acknowledged but has not adopted (*Report of the Conference of the Parties on Its Fifteenth Session,* held in Copenhagen December from 7 to 19 2009, *Addendum Part Two: Action Taken by the Conference of the Parties at Its Fifteenth Session,* UNFCCC, FCCC/CP/2009/11/Add.1, March 30, 2010). In 2010, the Obama administration pledged, under this Accord, that the United States would reduce greenhouse gas emissions by 17 percent below 2005 levels by 2020. (Lisa Friedman, "Nations Take First Steps on Copenhagen 'Accord,'" *New York Times,* January 29, 2010, http://www.nytimes.com/cwire/2010/01/29/29climatewire-nations-take-first-steps-on-copenhagen-accor-35621.html?pagewanted=all.

14. Todd Stern, Special U.S. Envoy for Climate Change, Remarks at Dartmouth College, August 2, 2012, http://www.state.gov/e/oes/rls/remarks/2012/196004.htm. ("The core assumption about how to address climate change is that you negotiate a treaty with binding emission targets stringent enough to meet a stipulated global goal . . . and that treaty in turn drives national action. . . . It makes perfect sense on paper. The trouble is it ignores the classic lesson that politics—including international politics—is the art of the possible.")

15. See Garrett Hardin, "Tragedy of the Commons," *Science* 162 (1968): 1243–1248.

16. European Commission. Emission Database for Global Atmospheric Research (EDGAR). http://edgar.jrc.ec.europa.eu Estimates of CO_2 emissions from fossil fuel combustion and some industrial processes.

17. Jody Freeman and Andrew Guzman, "Climate Change and U.S. Interests," *Columbia Law Review* 109 (2009): 1532; Kirsten H. Engel and Scott Saleska, "Subglobal Regulation of the Global Commons: The Case of Climate Change," 3 *Ecology Law Quarterly* 32 (2005): 183–233.

18. National Climate Program Act of 1978, 15 U.S.C. §§ 2901, et seq., 15 U.S.C. 2902. In the 1980 Energy Security Act, Congress directed that the National Academy of Sciences carry out a "comprehensive study of the projected impact on the level of carbon dioxide in the atmosphere, of fossil fuel combustion, coal-conversion and related synthetic fuels activities" authorized by the Act, and that the academy furthermore assess "the economic, physical, climatic and social effects of such impacts." Energy Security Act, 42 U.S.C. §§ 8701 et seq., 42 USC 8911 (1980). The 1987 Global Climate Protection Act directed EPA to propose to Congress a "coordinated national policy on global climate change" and ordered the secretary of state to work "through the channels of multilateral diplomacy" and "coordinate diplomatic efforts to combat global warming." Global Climate Protection Act, Title XI of Pub. L. 100–204, 101 Stat. 1407, note following 15 U.S.C. § 2901.

19. Global Change Research Act, 15 U.S.C. §§ 2931–2938.

20. 15 U.S.C. § 2933.

21. 15 U.S.C. § 2936(3).

22. Global Change Research Act, 15 U.S.C. §§ 2931–2938, § 2953.

23. Ibid.

24. These figures are found in Salvatore Lazzari, Congressional Research Service, *Energy Tax Policy: History and Current Issues,* CRS-11 (2008), http://www.fas.org/sgp/crs/misc/RL33578.pdf. Much higher subsidy figures have been claimed by leading energy-related public interest organizations. See, e.g., Public Citizen, *The Best Energy Bill Corporations Could Buy: Summary of Industry Giveaways in the 2005 Energy Bill* (Washington, DC: Public Citizen), http://www.citizen.org/documents/aug2005ebsum.pdf (estimating the bill provided $6 billion in oil and gas subsidies, $9 billion in coal subsidies, and $12 billion in subsidies to the nuclear power industry).

25. See *Summary of the Lieberman–McCain Climate Stewardship Act of 2003,* Center for Climate and Energy Solutions, http://www.c2es.org/federal/congress/108/summary-mccain-lieberman-climate-stewardship-act-2003.

26. Ibid.

27. See Ryan Lizza, "As the World Burns," *The New Yorker* (October 11, 2010): 11.

28. See Theda Skocpol, *Naming the Problem: What It Will Take to Counter Extremism and Engage Americans in the Fights Against Global Warming,* prepared for the Symposium on The Politics of America's Fight Against Global Warming, Harvard University, Cambridge, MA (February 14, 2013), 55.

29. Barack Obama, State of the Union address, February 12, 2013.

30. White House, *President Obama's Plan to Fight Climate Change,* http://www.whitehouse.gov/share/climate-action-plan.

31. Restatement (Second) of Torts § 821B (1977).

32. Complaint for Damages and Declaratory Judgment at 2, California ex rel. Lockyer v. Gen. Motors Corp., No. 06–05755 (N.D. Cal. Sept. 20, 2006).

33. Complaint at 43–49, Connecticut v. Am. Elec. Power Co., 406 F.Supp. 2d 265 (S.D.N.Y. 2005) (No. 04–5669).

34. Complaint, Kivalina v. ExxonMobil Co., No. CV-08–1138 (N.D. Cal. filed Feb. 26, 2008).

35. Connecticut v. AEP, 406 F.Supp. 2d 265, 274 (S.D.N.Y. 2005), vacated, 582 F.3d 309 (2d Cir. 2009), rev'd, 131 S. Ct. 2527 (2011).

36. Ibid., 273.

37. 582 F.3d 332 (2d. Cir. 2009).

38. Connecticut v. AEP, 582 F.3d 309, 325 (2d Cir. 2009), rev'd, 131 S. Ct. 2527 (2011).

39. David A. Dana, "The Mismatch Between Public Nuisance and Global Warming," *Supreme Court Economic Review* 18, no. 9, (2010): 9–42, 12.

40. American Electric Power Co. v. Connecticut, 131 S. Ct. 2527, 2538 (2011).

41. Village of Kivalina v. Exxon Mobil, 696 F.3d 849 (9th Cir. 2012).

42. See Doug Obey, "Backers of C02 Curbs Eye Liability Relief to Bolster Industry Support," *InsideEPA.com* (September 8, 2006). ("Capitol Hill supporters of legislation to regulate carbon dioxide [CO_2] emissions are quietly considering whether to provide industry liability protection from global warming–related lawsuits, in an effort to win backing for mandatory GHG limits, according to Capitol Hill and other sources.")

43. Petition for Rulemaking and Collateral Relief Seeking the Regulation of Greenhouse Gas Emissions From New Motor Vehicles Under §202 of the Clean Air Act (1999).
44. 42 U.S.C. § 7521 (2012).
45. US Department of Transportation, *Transportation's Role in Reducing U.S. Greenhouse Gas Emissions: Vol. 1, Synthesis Report.* Report to Congress. Washington, DC: US Department of Transportation (2010), 2–7, http://ntl.bts.gov/lib/32000/32700/32779/DOT_Climate_Change_Report_-_April_2010_-_Volume_1_and_2.pdf.
46. Memorandum from Robert E. Fabricant, general counsel, to Marianne L. Horinko, acting administrator, "EPA's Authority to Impose Mandatory Controls to Address Global Climate Change Under the Clean Air Act" (August 23, 2003), 2.
47. Ibid., 3.
48. 68 Fed. Reg. 52922 (2003).
49. Ibid.
50. Massachusetts v. EPA, 549 U.S. 497, 523 (2007).
51. Massachusetts v. EPA, 549 U.S. 497, 524 (2007).
52. Ibid., 529 (stating, "The statute is unambiguous").
53. 74 Fed. Reg. 66496 (2009).
54. EPA finalized emission standards for light-duty vehicles (2012–2016 model years) in May of 2010 and heavy-duty vehicles (2014–2018 model years) in August of 2011. These standards were later extended to years 2017–2025. 77 Fed. Reg. 62623 (2012).
55. EPA, "Standards of Performance for Greenhouse Gas Emissions for New Stationary Sources: Electric Utility Generating Units," 77 Fed. Reg. 22392 (2012).
56. 42 USC §§ 7470–7492 (2012).
57. See Michael B. Gerrard, "Climate Change Litigation in the United States," http://www.climatecasechart.com.
58. Center for Biological Diversity v. U.S. EPA, No. 2:09-cv-00670-JCC (W.D. Wash. 2009).
59. Center for Biological Diversity v. National Highway Traffic Safety Administration, 538 F.3d 1172, 1213 (9th Cir. 2008).
60. Border Power Plant Working Group v. Dept. of Energy, 260 F.Supp. 2d 997 (S.D. Cal. 2003).
61. Mid States Coalition for Progress v. Surface Transportation Board, 345 F.3d 520, 532 (8th Cir. 2003).
62. Ibid., 549–550.
63. Ibid., 550.
64. Center for Biological Diversity, 538 F.3d at 1217.
65. Robin Bravender, "16 'Endangerment' Lawsuits Filed Against EPA Before Deadline," *New York Times,* February 17, 2010, http://www.nytimes.com/gwire/2010/02/17/17greenwire-16-endangerment-lawsuits-filed-against-epa-bef-74640.html.
66. "Prevention of Significant Deterioration (PSD) and Tailoring Rule," 75 Fed. Reg. 31514, 31539 (2010) (estimating the need of a total of almost 10,000 new FTEs to process permits annually under one of the permit programs and 458 million additional work hours, and $21 billion in annual additional costs under the second permit program).
67. Coalition for Responsible Regulation v. EPA, No. 09–1322 (D.C. Cir. 2012).

CHAPTER 3

A New Era in States' Climate Policies?

Barry G. Rabe

THE UNITED STATES HAS BEEN widely condemned for its repeated inability to forge bold national strategies to address climate change. This opprobrium targets the limited capacity of federal governing institutions to devise policies to stabilize and reduce American greenhouse gas emissions, as well as their inability to assume a credible leadership role in international treaty deliberations. Although not a new issue, climate change remains a perplexing one for American federal institutions such as the presidency, Congress, and executive agencies. Ironically, the one branch of the American federal government that has had few inhibitions about major engagement is the judiciary, beginning with the historic 2007 *Massachusetts v. EPA* decision. This case was brought by a set of a dozen states that forced the federal executive branch to consider formally classifying greenhouse gases as air pollutants under federal clean air legislation. But even American federal courts have limited power to force policy formation and implementation, as Kirsten H. Engel notes in the previous chapter, leaving considerable questions about any future American national commitments on this issue. Indeed, federal inertia has had the largely unanticipated effect of shifting the locus of most climate-related policy development to subfederal levels, producing a patchwork quilt of state and local government policies. Even those policies adopted by the federal government have been, as we shall see, heavily reliant on states for either initial policy development or central roles in implementation, leading to a remarkably decentralized governance approach for an issue generally framed as a "global" problem.

The expansive state role has emerged over several decades, though it was largely unanticipated by scholars and policy makers, who assumed that only national and international institutions could design and implement climate policy. But American states, working independently or in collaboration with each other, may well have climatic, economic, and political incentives to take unilateral actions prior to federal and international engagement.[1] These may include unique and localized impacts of early evidence of climate change, with a desire to begin to mitigate potential local effects and also prepare for adaptation strategies. They may also reflect economic considerations, as states see investment in clean energy technologies as attractive venues for economic development, possibly positioning them to be national or international leaders in development of skills and technologies likely to be in demand in a decarbonized world. And early state action may also hold considerable political appeal, giving individual states and their state policy entrepreneurs national visibility, positioning them for influence in future federal policy design.[2] All of these factors likely contributed to the fact that American climate policy looks considerably more robust when one moves from the federal to the state level, as is increasingly evident in other federal governing systems, such as those of Australia, Canada, and India.[3] State governments continue to be prominent players, even after some new federal policies have gone into operation and some states have reversed earlier policy commitments.

This chapter will revisit the American case, with particular attention to the evolving role of state governments in policy design and implementation. It will acknowledge numerous federal-level impediments, building on Kirsten H. Engel's chapter, but note a continuing pattern of state engagement. Consequently, the chapter will contend that the United States *does* have a "climate policy," albeit one that consists of a number of rather fragmented pieces rather than a single, comprehensive initiative. If one views these various state and federal components collectively, and also considers lessons from the urban arena as discussed in Rachel M. Krause's chapter, a somewhat different story emerges from the conventional depiction of the United States as a pure laggard. Instead, there may be significant potential for emissions reduction through implementation of these various policies in the coming decades. These could build on the unanticipated stabilization and then decline in American emissions that has occurred over the past several years, though the continuation of this pattern is highly uncertain. Recent factors driving down emissions have included the economic collapse and related decline in energy demand, substantial replacement of coal with natural gas given shale drilling advances, and subfederal climate policy implementation. States may actually be poised for an expanded role in coming years, as they consider future energy and economic development options and also respond to evolving federal government policies.

The Collapse of National American Climate Policy Development

Climate change is hardly the first area that has defied development of a comprehensive and seamless policy response by the federal government. The very fragmentation of federal institutions mitigates against integrative policy design, particularly with the proliferation of veto points in the legislative and executive branches and the penchant for divided partisan control over the majority of years in the past half-century.[4] In turn, America remains a highly decentralized polity, leaving enormous areas of jurisdiction to state governments, which have their own constitutions, political cultures, and governance structures. Indeed, one need only think of such fragmented and contentious arenas as medical care and education policy to comprehend the enduring American inability to devise national policy that is politically sustainable, cost-effective, administratively feasible, and capable of meeting performance goals.

Climate change policy is nonetheless a distinctly difficult political challenge for American institutions, given its common framing as heavy on imposing front-loaded costs and uncertain at best on conferring long-term benefits. This has led to considerable effort to package climate policy proposals as shifting that equation and offering potential near-term benefits such as economic diversification and development. Such a reframing strategy reached its apex in November 2008: The election of the 44th president, a shift to Democratic Party control of both chambers of Congress, and a strong emphasis on potential cobenefits from reducing greenhouse gas emissions alongside facilitation of a far-reaching transformation of American energy policy seemed likely to result in the enactment of major legislation. At this point, it appeared probable that the federal government would assume the dominant role in American climate change policy, replacing state and local governments as policy leaders.[5]

The 111th Congress thus convened with considerable national and international expectations that it would produce comprehensive climate legislation that might well constrain or even formally preempt existing state and regional policies. This appeared likely to build on the momentum of predecessor Congresses that markedly expanded hearings on climate change and refined policy proposals that embraced emissions trading as the central feature of any new policy. Much of this deliberation drew liberally from the American experience in establishing a cap-and-trade program to address sulfur dioxide emissions from coal-burning power plants through 1990 clean air legislation supported by both a Republican president (George H. W. Bush) and a predominantly Democratic Congress.[6] In turn, coalitions of environmental groups and varied industry leaders began to converge to negotiate common ground and attempt to shape congressional decisions. At the same time, states began to position themselves individually and collectively for maximal advantage under an anticipated federal climate policy regime, possibly swapping their projected loss of authority over policy in exchange for shaping the terms of any new federal plan.

President Obama seemingly set the tone for a major American climate initiative in his first address to Congress, as he called in February 2009 for "this Congress to send me legislation that places a market-based cap on carbon pollution and drives the production of more renewable energy in America." The House of Representatives seemed particularly eager to oblige and four months later approved a 1,427-page bill that called for a 17 percent reduction in American emissions from 2005 levels by 2020. The American Clean Energy and Security Act, also known as "Waxman–Markey" for primary sponsors Henry Waxman (D-CA) and Edward Markey (D-MA), established a complex emissions cap-and-trade system and also added a wide range of other regulatory provisions and incentives in attempting to achieve those emission reduction targets. The legislation passed narrowly, on a 219-212 vote that largely split along party lines. But it appeared to be on a fast track for consideration and anticipated approval by the Senate, with the president promptly noting his general support for the bill and willingness to sign it or something like it. If it had been enacted, the legislation would have placed all existing state and regional cap-and-trade programs into a deep freeze through 2017, preventing them from operating during this period but holding out the option that they might be allowed to restart at a later date.

Rather than the beginning of a major policy development process, however, this was literally the beginning of the end of serious climate policy deliberations in the 111th Congress. Companion versions of Waxman–Markey were quickly introduced in the Senate, but none came close to a floor vote. Several key factors conspired to make the Senate particularly inhospitable to consideration of climate legislation in 2009 and 2010. First, the Senate's composition (two members per state) gave outsized influence to legislators from fossil fuel–dependent states, making the Senate's supermajority requirement to enact legislation (60 of 100 members) particularly formidable for any legislation that would threaten fossil fuel extraction or use. Second, the Great Recession hit with unexpectedly strong and extended force, giving the issue of economic recovery predominance over all other issues in American life and thereby marginalizing questions such as long-term climate change mitigation. Third, President Obama ultimately decided to invest his political capital not only in his economic recovery strategy but also a massive reform of the American health care system. The Patient Protection and Affordable Care Act was ultimately enacted and survived a Supreme Court challenge. But it proved extremely controversial and divisive and served to further push climate change to the recesses of Senate deliberations. Any further prospects for congressional engagement on climate change were dashed by the 2010 national elections, which shifted partisan control of the House of Representatives and led to substantial interbranch conflict with the president. This election brought to power a significant number of legislators in both chambers who questioned whether or not there was credible scientific evidence of climate change, much less need for any policy to address the phenomenon. In stunningly short order, Congress shifted from a hotbed of active discussion of climate policy options to a prominent stage for the American effort to discredit the very existence of climate change, thereby dashing earlier expectations of a comprehensive federal strategy. But none of this precluded states from sustaining, expanding, or abandoning previous policy commitments.

The New Normal: Implementing Bits and Pieces of a National Climate Strategy

There appears to be little if any prospect of revisiting the high-water mark of federal legislative exploration of climate policy in the next several years. Indeed, it remains difficult to envision any near-term scenario that would realistically prompt any future Congress into action on climate change, barring wrenching shifts in weather and storm activity that gave the issue greater salience. Nonetheless, it would be inaccurate to suggest that the United States moves ahead without any semblance of policy designed to reduce greenhouse gas emissions. A diverse set of policies appears to be heading into advanced stages of implementation. Most tend to place state governments in a central role moving forward, whether through policies of their own creation or through a lead role in implementing various federal initiatives. Few economists would embrace this mixture of policies as the most cost-effective approach, and yet there are abundant precedents in the United States for addressing policy issues through a hodgepodge of separate initiatives rather than one uniform and seamless strategy. Collectively, these American policies, if sustained, begin to compose an approach that could indeed continue the recent trend to stabilize and even reduce emissions.

In the best case, full implementation of this set of programs could ironically move the United States into an emissions reduction trajectory roughly in line with what was proposed under the 2009 legislation that passed the House. In perhaps the most thorough analysis of American climate policies to date, a team of analysts from the World Resources Institute assessed "federal regulatory scenarios and state scenarios" and offered alternative emissions reduction paths that might be followed. Using a 2005 emissions baseline, they concluded that "lackluster" implementation would produce a 6 percent reduction in emissions by 2020. But this climbed to 9 percent under a "middle-of-the-road" approach and jumped to 14 percent under a "go-getter" approach. Extending this analysis through 2030 produced a 27 percent reduction scenario from 2005 levels under the go-getter approach. All of this was based on existing federal and state policies but with the reductions contingent on the resiliency and intensity of implementation.[7] This assessment may prove conservative, given the unexpected rate of emissions decline in recent years, as well as emerging developments that will be discussed later in this chapter. Ironically, the United States may not need comprehensive federal legislation to begin to achieve some emissions reduction targets through a combination of other mechanisms.

The New Era of American Climate Governance

Federal engagement on a policy issue is not necessarily confined to Congress and the legislative process. Under the American Constitution, the executive branch reserves considerable powers of both policy initiation and interpretation. Indeed, a recurring theme in the study of the American executive branch is the continual use of "administrative presidency" powers, whereby a president can take significant unilateral steps when it proves impossible to reach agreement with Congress. In environmental policy, there is ample precedent for this approach, ranging from the formation of the EPA under Richard Nixon to reinterpretation of the Clean Air Act under George W. Bush that offered regulated parties greater compliance flexibility. Such an approach has been aggressively pursued in the Obama administration, accelerating after the 2010 and 2012 elections and applied with particular rigor in the area of climate policy. Ironically, these efforts to establish more active federal engagement not only are designed to bypass Congress but routinely place primary reliance upon states and their lead environmental agencies for interpretation and implementation.

Revisiting the Clean Air Act

While the 1990 Clean Air Act is best remembered for ushering in emissions trading for sulfur dioxide under Title IV, it also included under Title V considerable tightening of conventional regulatory standards for many other major point sources of air pollution. The legislation also provided some flexibility, whereby the executive branch and the Environmental Protection Agency could make future adjustments as new scientific evidence emerged concerning risks from exposure to air contaminants. These possible adjustments included the addition of various air emissions that science found to pose a public health threat as well as revisiting regulatory standards over time. As is the case with most American environmental legislation, there has been no successor legislation to the 1990 law, reflecting the ongoing deadlock in respective Congresses and protracted conflicts with various presidents. This gave President Obama considerable latitude in reconfiguring an older statute to serve the new purpose of reducing greenhouse gas emissions, albeit one heavily dependent on state government interpretation and contingent on his ability to remain in office. The president embraced the 2007 Supreme Court decision discussed in the prior chapter (*Massachusetts v. EPA*), leading to a prompt EPA "endangerment finding" that deemed carbon dioxide to be an air pollutant, thereby subject to the terms of the Clean Air Act. Obama threatened to use this finding to expand federal regulatory power over carbon emissions if Congress failed to enact climate legislation. The 2010 collapse of Senate climate deliberations then prompted the president to begin applying clean air standards to new power plants in 2012, with an expansion of coverage to all operational plants embraced after his 2012 re-election.

Air quality permits are generally issued by state environmental agencies in conjunction with "state implementation plans" negotiated with the EPA. But states have very different philosophies and capacities, and the first years of intergovernmental implementation suggests that this process could lead to very different application of these provisions in different parts of the nation. A number of states have devised greenhouse gas emissions reduction programs in previous years; these generally tend to be "leader" states with strong environmental enforcement and performance records, including nearly all states in the Northeast and along the Pacific Coast. Many of these states have approached the new EPA requirements as an opportunity to gain credit for their own climate policy commitments and early emissions reductions, potentially easing federal compliance processes. There is considerable precedent in federal environmental policy and American intergovernmental management more generally to provide rewards and incentives for so-called early movers, in some cases encouraging them to "race to the top." States may have particular latitude in this case to think outside the box, as there is no singular best available control technology for reducing greenhouse gas emissions that can be uniformly applied to all sources, particularly as the federal policy expands to cover not only proposed plants but also established ones. Indeed, the 2013 confirmation of Gina McCarthy to head the EPA included strong signals that the agency wanted to make climate change a pillar of collaborative relationships with states that included tangible federal incentives for outstanding innovation and performance. McCarthy, who formerly oversaw environmental agencies in Connecticut and Massachusetts, designated "Launching a New Era of State, Tribal and Local Partnerships" as one of her top priorities and began an active outreach process with states.[8]

At the same time, not all states may see this as an opportunity for innovative environmental governance. In fact, the federal government may intensify oversight pressure on any states deemed laggards. Texas is the largest source of greenhouse gas emissions among the 50 states and has the largest number of industrial facilities likely to fall under Clean Air Act auspices. But the state has repeatedly and stridently rejected the notion that there is any legitimacy to the EPA climate effort. This has led to denunciations of the Obama administration effort by Governor Rick Perry and Attorney General Greg Abbott as well as a leadership role among states trying to thwart this process through litigation. At the same time, it may lead to a particularly contentious and expensive implementation process, as EPA staff began in 2012 to assume responsibility for this permitting process from the Texas Commission on Environmental Quality, given the refusal of state authorities to cooperate. It appeared that this expanded federal role might foster a more rigorous application of federal permit provisions in Texas than in other states, producing concern among some regulated parties that the state's intransigence may ultimately prove costly for them. State responses to this emerging form of federal engagement will be a significant test of their commitment to finding innovative and effective ways to reduce greenhouse gas emissions.

Vehicular Fuel Efficiency

Presidents have long retained authority over altering regulatory standards for fuel efficiency levels in new cars and trucks. This stems from energy legislation enacted in the mid-1970s, although most presidents have moved cautiously or entirely ignored opportunities to tighten standards. The 2007 Energy Independence and Security Act reopened this issue, establishing a slight increase in fuel efficiency standards but delegating future decisions to the executive branch. As in the case of air emissions, the Obama administration has used these powers aggressively. This began with a 2009 agreement that required a five-year phasein of major fuel economy increases through 2016, from 27 miles per gallon in 2011 to 35.4 miles per gallon in 2016.

This followed a period of intensive intergovernmental bargaining, with the administration in essence embracing a legislative proposal from California and allied states to take an ambitious stance on this issue. California reserves unique status on federal legislation to establish air quality standards for vehicles above the levels of existing federal standards. If approved by the federal government through a "waiver" process, any other state can then establish standards at the same level as those of California.[9] This has frequently generated a ratcheting-up effect, whereby California acts first, some other states join it in alliance, and the federal government embraces the position as a national standard to create a uniform system. President Obama in fact announced the 2009 agreement at a Rose Garden ceremony featuring three prominent governors.

But this was only the beginning of an expanded administrative presidency stance on this issue. In November 2011, President Obama announced that the 2016 fuel efficiency targets would be the beginning and not the end of the expanded use of this policy tool. As of 2017, fuel efficiency would be required to increase by 5 percent each year for cars, and between 3.5 and 5 percent annually for light trucks, reaching a level of 54.5 miles per gallon by 2025. This would effectively double current levels of fuel efficiency by the middle of the next decade. As with EPA air regulatory standards, this has proven controversial and triggered concerns in Congress. But it also demonstrates the role of individual states as a potential lever for federal action, given the pivotal role of California in prodding a national response.

Future Prospects

These federal initiatives all demonstrate the authority of a federally elected executive to work with existing statutes but markedly expand their scope. They could have significant consequences on American greenhouse gas emissions for decades to come. Any future impact, of course, will be contingent on continuing executive branch support for implementation, though this appeared likely in the aftermath of President Obama's 2012 re-election and selection of a new EPA leadership team to sustain this approach. Moreover, the EPA's new use of air quality legislation is heavily contingent on some form of administrative collaboration with states issuing permits through their state implementation plans. In turn, states present other possibilities for additional greenhouse gas emission reductions through their own unilateral or collective efforts, continuing a prominent role that they began to assume in the 1990s.

The Evolving Era of State Climate Governance

State governments have continued to adjust their role in climate policy. During an extended period of federal inertia in previous decades, many states launched unilateral policy experiments, with strong emphases on renewable energy development and promotion of energy efficiency. States face some constitutional constraints on policy options, most notably those imposed by the commerce clause, which precludes any restriction on interstate movement of goods and services. But a vast array of state policies were enacted during a period of state domination of American climate policy that ran from the late 1990s through the latter 2000s, when a significant federal role seemed politically infeasible.[10] Many of these policies are positioned to move into advanced stages of implementation, with potentially significant impacts on emissions.

However, the state government role in climate policy has faced a series of challenges in recent years, generating questions about the resiliency of policies enacted in recent decades. First, the pace of state climate policy development and the diffusion of innovations across multiple states slowed dramatically in the late 2000s. This was attributable in part to the growing expectation that the federal government was likely to enact a comprehensive climate policy and thereby assume a dominant role. Under many competing federal policy options, some form of preemption of existing state policies was prominent, as discussed above in the case of the Waxman–Markey legislation. States slowed their initiation of new policies amid this uncertainty and instead began to position themselves for maximal advantage under any new federal policy regime. They have more recently begun to readjust to the reduced likelihood of major federal intervention at any point in the near future while also preparing to assume a lead role in air quality permit development.

Second, some state climate policies have faced a political backlash and possible repeal or retrenchment. This reflects some major shifts in state government leadership, particularly key gubernatorial transitions. Executive branch swings in states such as Arizona, New Mexico, and Utah contributed to the collapse in 2011 of the Western Climate Initiative (WCI). The WCI had begun several years earlier as a partnership between seven states and four provinces with the goal of establishing a cap-and-trade program that sought a 15 percent reduction in all greenhouse gas emissions from 2005 levels by 2020. Provincial commitment has also waned, leaving California and Quebec as the remaining partners to begin a 2013 launch of carbon cap-and-trade, though other jurisdictions have indicated some interest in revisiting this or pursuing other cross-unit partnerships.

A similar initiative among midwestern states and the province of Manitoba, known as the Midwestern Greenhouse Gas Reduction Accord (MGGRA), suffered a similar implosion, with a 2007 memorandum of understanding now essentially ignored by all participating jurisdictions. Even the original regional cap-and-trade program, the Regional Greenhouse Gas Initiative (RGGI), faced challenges as it moved into its sixth year of operations in 2013. Most notably, New Jersey formally withdrew from the interstate program in 2011, and a few others began to contemplate a similar step. However, RGGI has maintained quarterly carbon allowance auctions, generating substantial funds for state renewable energy and energy efficiency programs, and all remaining states tightened their carbon caps by more than 30 percent in 2013. Other state-specific climate policies moved ahead while also facing implementation challenges. For example, states such as Colorado, Connecticut, Minnesota, and Montana considered scaling back their plans to expand renewable energy through state mandates known as portfolio standards. Such standards mandate a steady increase in the supply of a state's electricity that comes from renewable sources. Opposition to these standards has included proposals to either delay implementation or reduce the levels of required new renewable energy development, although no major changes had been approved as of mid-2013. Thirty-six states retained climate action plans that estimated greenhouse gas emissions and outlined possible policy steps, but most of these have lacked statutory teeth and have frequently proven easy for state governments to ignore. Questions have arisen as to whether or not these plans had produced enduring policies or demonstrable emission reductions or were instead largely analytical and symbolic exercises.[11] The vast majority of state policies enacted in recent decades remain operational, but this reversal on some high-saliency cases raises questions about their long-term sustainability as political leaders and economic conditions change.

Third, a cluster of states shifted from positions of neutrality or indifference to climate policy toward active opposition toward emerging federal policies. The governor and attorney general of Texas have not only opposed new EPA initiatives regarding greenhouse gas and air emission permits but have led multistate efforts to challenge these in court. They have also emerged as outspoken opponents of new vehicular fuel efficiency requirements and renewable fuel standards reliant on corn-based ethanol. But Rick Perry and Greg Abbott have increasingly had company in this arena, with at least 16 states joining one or more of their legal challenges. States attorneys general have been particularly active in exploring ways to challenge the legitimacy of federal action as well as climate science. Virginia's Mark Cuccinelli, for example, routinely expressed his doubts about the existence of climate change and actively opposed many Obama administration policy interpretations. He also sought release of documents from the University of Virginia concerning the climate research of a former faculty member, alleging that this might reveal instances of fraudulent use of state funds. Although states attorneys general are elected on a partisan basis, state environmental agency heads are usually appointed by their governors and tend to follow closely the directions of the elected executive.

These developments have served to challenge but not eviscerate the ongoing state government role in climate policy. Alongside these emerging stumbling blocks, a substantial body of policy implementation continues to move forward. In many instances, states have reaffirmed their policy commitments or even expanded them in various ways. This leaves an uneven pattern of state policy engagement, with the states most intensively involved generally located along either the west or east coasts. But they serve to sustain the "bottom-up" element of American climate policy development that emerged so unexpectedly in the 1990s and 2000s and is linked in many ways to evolving federal efforts.

The Dominant Early Mover: California

California has long emerged in studies of American state government as among the very first to take unilateral steps in emerging areas of public policy. Environmental protection has long been among those areas of particularly active interest, including development of an aggressive air quality regime well in advance of the federal government.[12] State air quality policies have contributed to a significant decline in conventional air contaminants and also served to give California the highest level of per capita energy efficiency in the nation. This early and sustained engagement also has given the state a position of unique influence in shaping federal policy.

In the clean air case, California's very early commitment to far-reaching policy enabled it to secure a unique agreement with Congress, allowing it to request a federal waiver to establish air quality standards above any federal baseline. As noted above in the vehicle emissions case, once federal approval is granted, which has occurred routinely across five decades, any other state may emulate the California policy. This creates the possibility of two competing standards in operation within different sets of states. It has regularly served as a prod for upward movement in national air quality standards, with the federal government frequently embracing the position originally taken in Sacramento. This was exactly the formula that led to major recent increases in federal vehicular fuel economy standards. California legislation enacted in 2002 (Assembly Bill 1493) recognized the lack of state jurisdiction to set fuel economy standards but instead set vehicular emission levels so as to coincide with preferred state fuel economy targets. This ultimately served as the lever for Obama administration decisions to make the California standard the national one.

The state has also sought aggressive unilateral reductions in greenhouse gas emissions and attempted to work collaboratively to the extent possible with other states and provinces. California's 2006 Global Warming Solutions Act (Assembly Bill 32) remains the most ambitious climate legislation enacted anywhere in North America and among the most aggressive policies in the world. The legislation established greenhouse gas reduction targets 15 percent below 2005 levels by 2020 and 80 percent below 1990 levels by 2050. It also ushered in a cap-and-trade system across multiple greenhouse gases and sectors of the economy that was expected to address approximately 85 percent of total state emissions. This policy decisively survived a ballot proposition challenge that would have essentially halted implementation, with Prop 23 defeated by a 61-to-39 percent margin in November 2010.

This step allowed movement toward full implementation, featuring a plan to establish an emissions trading infrastructure, convene quarterly auctions, develop public disclosure and reporting provisions, and allow Quebec and possibly other states or provinces to join as partners. This was designed to allow transition into full operation in January 2013 through a series of regular carbon allowance auctions. These auctions were expected to generate more than $1 billion of new revenue for California per year, with considerable political support for using the bulk of these funds to promote energy efficiency and renewable energy or to support local communities facing the most serious immediate impacts of climate change. Several court challenges remain, however, including suits filed by oil refineries, environmental justice groups, and firms that generate electricity in other states. Given the implosion of the WCI network, California now represents the most prominent North American experiment in attempting carbon cap-and-trade on a large scale, with an enduring base of political support and growing interest in forming partnerships beyond North America.

California's approach to climate change goes well beyond fuel economy standards and carbon cap-and-trade, however. At the same time that some states are considering easing their renewable portfolio standard (RPS) targets, California has moved in the opposite direction. Legislation enacted in April 2011 increased the state's binding renewables target for 2020 from 20 percent to 33 percent, with interim targets of 20 percent by the end of 2013 and 25 percent by the end of 2016. The state is also working to include more than 40 publicly owned utilities in the RPS process, thereby eliminating their earlier exemption. The state also continues to pursue a wide range of energy efficiency programs, a low-carbon fuel standard, feed-in-tariffs for small renewable energy projects, and a public benefit charge on electricity use that helps fund many of these state programs. The low-carbon fuel standard, however, has been challenged in court, as other states contend California is trying to impose its policies beyond state borders, thereby raising possible violation of the commerce clause.[13] In turn, the benefit charge faces some uncertainty, since its reauthorization would require a supermajority vote given 2010 changes in the California constitution that classify some fees and charges as, in essence, taxes.

Collectively, these various programs serve to maintain California's status as an American early mover on climate change. Former Governor Arnold Schwarzenegger and other state leaders laid claim to state status as a world leader on climate change in 2006, when the Global Warming Solutions Act was signed into law. California further promoted itself as developing the "first-in-the-world comprehensive program" to combat climate change as well as "the most radical climate policy in the world."[14] At the time, this seemed a bit hyperbolic and, if anything, likely to soon be overtaken by federal initiatives in Washington DC and Ottawa, among other capitals. But the basic infrastructure of this approach has endured leadership changes as well as both political and legal challenges, leaving California in a unique position nationally, continentally, and globally. California political and climate policy leaders would no doubt welcome more state and provincial allies in their efforts, including states such as Arizona, Oregon, and Washington and provinces such as British Columbia, Manitoba, and Ontario that made bold initial pledges for collaboration through the WCI but have largely backtracked as implementation neared. Nonetheless, Quebec has remained a fully active partner, and other jurisdictions may revisit this, in part because of new federal incentives to take constructive policy steps.

Energy Efficiency Resource Standards

California may remain an outlier among states, although it hardly stands alone in trying to find ways to reduce greenhouse gas emissions. One policy tool that has received growing political support and experienced increased state adoption is an energy efficiency resource standard (EERS), a "performance-based mechanism that requires electricity and natural gas distributors to achieve a percentage of energy savings relative to a baseline."[15] Twenty-four states have now established some version of an EERS, either on a stand-alone basis (such as Arizona, Indiana, Ohio, and New Mexico) or embedded into an RPS (such as Connecticut, Nevada, and North Carolina). According to an October 2011 report by the American Council for an Energy-Efficient Economy (ACEEE), states have continued to expand not only energy efficiency resource standards, but also a range of related energy efficiency initiatives, such as building codes, electricity decoupling programs, public benefit charge programs to fund expanded energy efficiency funding, and government purchase of high-efficiency vehicles and lights. In some cases, this entails coming into full compliance with either professional codes of conduct or international best practice standards, such as the International Energy Conservation Code. According to the ACEEE report, "Amid the acrimonious debates over state budget deficits, state government policymakers from both sides of the aisle pushed for energy efficiency in homes, businesses, and their own state government facilities."[16] Many of these state policies could clearly be coordinated with those of neighboring states, including state–provincial collaboration that would build on existing state and provincial partnerships in electricity and other arenas of energy policy. As Linda Breggin of the Environmental Law Institute has noted, "What is remarkable is not that California is leading the country but how many other states are on the move as well."[17] The ACEEE has produced a comprehensive ranking of the 50 states based on the intensity of their commitment across various policy options. As noted in Table 3.1, such states as Massachusetts, California, New York, Oregon, Connecticut, Rhode Island, and Vermont have received the highest scores in recent years.

TABLE 3.1   State Energy Efficiency Rankings, 2013

Rank	State	Score* 2013	Score 2012
1	Massachusetts	42	43.5
2	California	41	40.5
3	New York	38	39
4	Oregon	37	37.5
5	Connecticut	36	34.5
6	Rhode Island	35.5	33
7	Vermont	34.5	35.5
8	Washington	33.5	32
9	Maryland	27.5	30
10	Illinois	26	25
11	Minnesota	25.5	30
12	New Jersey	24.5	24.5
12	Arizona	24.5	25.5
12	Michigan	24.5	25.5
12	Iowa	24.5	26.5
16	Maine	23	19
16	Colorado	23	25
18	Ohio	22.5	19.5
19	Pennsylvania	22	21.5
20	Hawaii	20.5	22
21	New Hampshire	20	22

Rank	State	Score 2013	Score 2012
22	Delaware	18.5	18.5
23	Wisconsin	18	22.5
24	New Mexico	17.5	18.5
24	North Carolina	17.5	19.5
24	Utah	17.5	20
27	Indiana	15.5	14
27	Florida	15.5	17.5
29	Montana	15	19
30	District of Columbia	14	17.5
31	Tennessee	13.5	15
31	Idaho	13.5	19.5
33	Georgia	13	14
33	Texas	13	14
33	Nevada	13	16.5
36	Virginia	12.5	13
37	Oklahoma	12	11
37	Arkansas	12	13
39	Kansas	11.5	8.5
39	Alabama	11.5	10.5
39	South Carolina	11.5	10.5
39	Kentucky	11.5	13.5
43	Missouri	10.5	9
44	Louisiana	9.5	9
44	Nebraska	9.5	9.5
46	West Virginia	9	6
47	Mississippi	8	2.5
47	Alaska	8	8
47	South Dakota	8	8
50	Wyoming	5.5	6.5
51	North Dakota	3.5	4

*The score is the combined total of points from each of the following categories, with 50 total possible points: utility and public benefits fund efficiency programs and policies score—20; transportation score—9; building energy code score—7; combined heat and power score—5; state government initiatives score—7; appliance efficiency standards score—2.

Source: American Council for Energy-Efficient Economy (2013).

Renewable Portfolio Standards

Despite some legislative proposals to set more modest targets in some states, the vast majority of the 29 states that have enacted RPSes are moving into advanced stages of implementation with solid political support. These RPSes operate in one or more states in every region of the United States except the Southeast, as depicted in Figure 3.1. Much like energy efficiency resource standards, RPSes have continued to retain considerable bipartisan support. Some states continue to embrace the RPS as a climate policy tool, whereas others focus primarily on other perceived RPS benefits, such as reduction of other environmental threats and possible economic development benefits. Collectively, RPSes remain a significant component of state climate strategies.

Ironically, Texas's outsized opposition to most federal efforts to reduce greenhouse gas emissions did not transfer to the issue of state RPS development. The first Texas RPS was signed into law in 1999 by then-Governor George W. Bush, and his successor, incumbent Rick Perry, signed a major expansion into law six years later. The level of electricity that Texas receives from renewable sources jumped from less than 1 percent in 2000 to more than 10 percent by 2013; this pattern of growth is likely to continue, as the role of wind is expected to expand in the state, while the role of coal will likely decline, with particularly robust wind

FIGURE 3.1 Renewable Portfolio Standards Policies

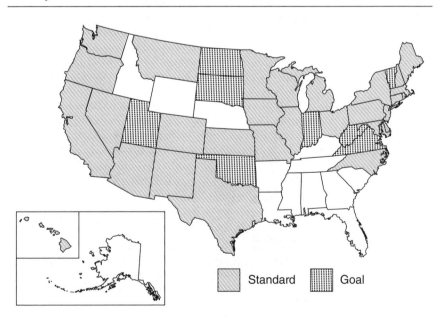

Standard   Goal

turbine expansion in the western part of the state. As Perry has noted, "Texas is the nation's leader in wind energy, thanks to our long-term commitment to bolstering renewable energy sources and diversifying the state's energy portfolio."[18]

Collectively, state RPSes are scheduled to add 76,650 megawatts of new renewable electricity by 2020, which would represent a 570 percent increase from 2000 levels. The majority of this new energy has been derived from wind, although a number of states are attempting to promote a wider mix of new sources, including solar and geothermal, in later rounds of RPS implementation. These policies continue to receive substantial public support in public opinion surveys, far more so than market-based tools such as cap-and-trade and carbon taxes.[19] They also create considerable opportunities for multijurisdictional partnerships given the cross-border movement of much electricity in Canada and the United States. However, the pace of expansion for these policies to additional states began to slow in the 2010s. As Diana Forster and Daniel A. Smith discuss in detail in Chapter 6, Michigan voters decisively rejected in November 2012 a proposed ballot proposition to expand their existing RPS to a target of 25 percent renewable energy by 2025. Interpretation of this vote was complicated by the decision to make this proposal a constitutional amendment, one of five such amendments on the ballot at the same time, all of which were defeated.

Coal Phaseout

The state RPS push to expand the base of renewable electricity sources potentially poses a challenge to providers of traditional energy sources, such as coal. But a series of state policies that have raised sobering questions about the desirability of approving permits for proposed new coal-burning facilities may also be contributing to a transition away from coal. Indeed, some of these steps may ultimately serve to expand natural gas use as well as work hand in hand with federal efforts to reduce conventional and greenhouse emissions from coal facilities, and thereby reduce coal's historically dominant role in electricity generation in the United States.

These state policies have taken different forms, often linked to formal procedures for approval of either new energy-generating facilities or expansion of existing ones. In Kansas, for example, the administration of former governor Kathleen Sebelius raised major concerns in 2007–08 about a proposal by the Sunflower Electric Power Corporation to open a pair of major coal-burning facilities that would produce 1,400 megawatts of electricity. State concerns focused on a mixture of issues, including conventional air contaminants, greenhouse gas releases, and whether such an expansion of electricity supply was needed in the state. This led to a decision to trim the proposal to a single facility that would produce 895 megawatts in conjunction with creation of an RPS for the state. It is possible that the new coal capacity will never be added, given declining demand and the emergence of alternatives that include natural gas.

Similarly, in Michigan, a 2009 executive order by then-Governor Jennifer Granholm required a thorough review of alternatives prior to regulatory approval of new coal-based facilities. This led to a state permit denial for a proposed new coal plant and then litigation. However, the utility proposing the facility withdrew its request in late 2011, citing changing market conditions. Alongside these more indirect attempts to block coal, other kinds of state measures have been enacted. One of the most aggressive is the 2009 Colorado New State Act, which is designed to facilitate a coal phaseout in favor of expanded reliance on natural gas and renewables in conjunction with its RPS. The pace of cancelling new coal-burning plants only accelerated in recent years, with one or more proposed plants shelved in such states as Georgia, Illinois, Mississippi, and Texas.

Resilient Regionalism and Implicit Carbon Pricing

The demise of both the WCI and its midwestern counterpart raise sobering questions about the sustainability of any multijurisdictional climate policy initiative. States and provinces have periodically signed regional agreements that are bold but nonbinding and hence largely symbolic. But these regional cap-and-trade pacts appeared to be quite different, reflecting more formal commitments by participants and extensive negotiations to facilitate implementation.[20] Even the RGGI has been weakened by the exit of New Jersey and the continued inability to secure participating pledges from other states and provinces that have instead retained nonbinding "observer" status.

The RGGI, nonetheless, continues to operate in nine northeastern states, running a cap-and-trade program designed to stabilize and gradually reduce carbon emissions from coal-burning power plants in the region.[21] The RGGI likely had little if any impact on overall emissions in its early years, following a rapid plunge in electricity demand in the region due largely to the recession that produced a rate of carbon emissions well below the current cap. At the same time, the RGGI has demonstrated that a carbon cap-and-trade system can run with what appears to be a high level of proficiency and transparency. Moreover, its use of the auctioning process to allocate allowances has generated significant revenue for participating states. Its first four years of quarterly auctions produced more than $1 billion, about 80 percent of which was transferred directly to support state energy efficiency and renewable energy programs. Indeed, this approach has had considerable influence on other jurisdictions committed to some form of carbon trading, including California and the European Union Emissions Trading Scheme, in embracing some form of auctioning rather than free allocation. Moreover, the nine RGGI states agreed to a significant lowering of the regional cap in 2013, producing an immediate increase in auction prices and likely future emission reductions.

Related to this, the RGGI auctions have emerged as one of a series of measures whereby states impose some form of tax, fee, or charge on the use of fossil fuels. This falls far short of a pure carbon tax but nonetheless has the effect of elevating the cost of using fossil fuels and then shifting those revenues directly toward programs that provide alternatives. Many states are currently revisiting the option of using severance taxes on the extraction of fossil fuels, creating a potentially large revenue source and building on the previous experience of energy-endowed states such as Alaska, Texas, Oklahoma, and Wyoming. The significant expansion of natural gas derived from hydraulic fracturing is a particularly prominent target for such consideration, leading a growing number of states to consider either creating such a tax or expanding existing ones, with a possible transfer of revenues to climate-friendly programs. There is no singular pattern yet emerging in this area, but these varied efforts reflect state capacity to set an "implicit carbon price" that, in effect, has the same effect of explicit carbon taxes in jurisdictions such as British Columbia in Canada and several European nations.[22]

Where Next for American Climate Policy? Emerging Issues

American politics and governance remain volatile, making any projection of future developments difficult at best. The Great Recession and uncertain recovery continued to dominate American political discourse in the 2010s. Climate change policy appears to be a relatively low priority for Congress and many states, although this could conceivably change in the coming years. Nonetheless, as this chapter has noted, a number of federal and state policy developments have converged to give the United States a climate policy strategy, albeit a patchwork process at considerable variance from earlier proposals for a comprehensive national system. Looking forward, there are several emerging factors that could influence the future shape of American policy and the national trend in greenhouse gas emissions.

Whither the Economy?

Perhaps the greatest climate irony of the past half-decade is that the collapse of federal negotiations over a far-reaching climate bill has coincided with an unanticipated plunge in American greenhouse gas emissions. According to a 2011 EPA study, national greenhouse gas emissions dropped approximately 7.7 percent between 2005 and 2009. There were outright declines in three of these years, with the steepest drops in emissions registered in 2008 and 2009. The EPA has noted that a primary contributor to this trend has been the decline in economic output that has decreased energy consumption across sectors. The agency also discovered a reduction in the carbon intensity of fuels that are used to supply energy; it is plausible that some combination of federal and state policies contributed to this transition, although the EPA has not estimated their impact to date. But clearly the American Great Recession triggered a level of emissions reduction that moved it during 2005 and 2009 approximately one-half of the distance established in the Waxman–Markey legislation through 2020 (15 percent below 2005 levels). Early estimates from 2010–12 suggest that there was only modest emissions rebound amid the early signs of recovery, perhaps suggesting a genuine shift toward somewhat greater energy efficiency. Indeed, the think tank Resources for the Future noted in late 2012 that "the expectation is that our emissions growth ahead will be modest."[23] When combined with the evolving American policy initiatives discussed above, it is possible to envision an American emissions trajectory over the coming decade that reflects stabilization or perhaps further decline, raising a question about the future impacts of various state and intergovernmental policies over the longer term.

Whither Fracking?

A surge in production of natural gas derived from hydraulic fracturing (or "fracking") techniques is a primary factor driving the transition from coal to this substantially cleaner fossil fuel source in American electricity. This represents a dramatic shift from much of the 2000s, when energy policy analysts anticipated a decline in American natural gas yields and soaring prices amid scarcity, thus making both coal and other alternatives appear more promising for the future. Natural gas produces only one-half of the carbon dioxide emissions per unit of energy generated that coal produces. According to a 2012 Resources for the Future analysis, "Carbon dioxide emissions are in decline not only as a result of the economic slowdown but also because of heightened efficiency and a change in our fuel mix, especially in the electricity sector."[24]

The absence of a federal regulatory regime for fracking leaves regulatory authority with the states, thus far producing a highly uneven pattern of policies.[25] Fracking poses a wide range of environmental concerns, including groundwater contamination, air emissions that can include greenhouse gases such as methane, management of wastewater that returns to the surface after use, and even seismic activity following chemical injection below ground. Despite these concerns, most states appear likely to encourage the expansion of fracking and thereby achieve a dramatic increase in national use of natural gas in the coming decades, with potentially large reductions in greenhouse gas emissions when coal is being replaced. Indeed, several states, such as Pennsylvania and Texas, saw significant electricity sector emission declines alongside substantial replacement of coal with natural gas in 2012.

This issue, however, remains highly contentious. On the one hand, natural gas could further marginalize coal use in electricity generation, and it could replace oil and gasoline as transportation fuels as well, should proposals to expand its use move forward. The rapid expansion of American supplies of both gas and oil from shale deposits also served to heighten the uncertainty surrounding proposals to expand exports of oil from Albertan tar sands via controversial new routes, such as the proposed Keystone XL pipeline. On the other hand, it is possible that natural gas will offer an available and inexpensive alternative to renewable sources such as wind and geothermal. The massive shale gas deposits scattered across the United States, as well as the rapid development of natural gas production recently, suggest that natural gas will be a far greater player in the American context than could have been envisioned just a few years ago, with states likely assuming a dominant role of policy development and formation in coming years.

Whither Public Opinion?

Public opinion on climate change, as discussed by Dennis Chong in Chapter 4, has undergone significant shifts in recent years. But there were numerous signs that public opinion began to "rebound" on this issue in 2011–12, reflecting increased measures of belief in climate change and support for some mitigation policies. The National Surveys of Energy and Environment at the University of Michigan and Muhlenberg College have been tracking these trends for a number of years, and they found in their fall 2012 survey that strong majorities of Americans felt that federal, state, and local governments should assume either "a great deal of responsibility" or "some responsibility" for "taking actions" to reduce climate change, as demonstrated in Table 3.2.[26] In turn, these findings reflected some increase in support for cross-level action from prior years, as noted in Table 3.3. This same survey found strong public support for national policy options such as renewable portfolio standards and increased vehicular fuel efficiency. It also found, for the first time, a plurality of support for a national carbon tax. In each case, however, support declined when a steep price tag was added.

TABLE 3.2   Public Support for Governmental Responsibility to Address Climate Change, 2012

Responsibility for Reducing Global Warming, by Level of Government

Level of government	A great deal of responsibility	Some responsibility	No responsibility	Not sure
The federal government	51%	22%	21%	6%
State governments	44%	28%	22%	7%
Local governments	38%	30%	26%	7%

Source: Center for Local State, and Urban Policy, University of Michigan (2012).

TABLE 3.3   Public Support for Governmental Responsibility to Address
Climate Change, 2012

Percentage of Americans Who Believe That Government Has a Great Deal of
Responsibility for Taking Actions to Reduce Global Warming

Level of government	Fall 2008	Fall 2009	Fall 2010	Fall 2011	Fall 2012
The federal government	48%	53%	43%	42%	51%
State governments	34%	37%	35%	32%	44%
Local governments	26%	34%	29%	29%	38%

Source: Center for Local State, and Urban Policy, University of Michigan (2012).

Intergovernmental Collaboration?

The collapse of the western and midwestern regional climate initiatives underscores the challenge of sustaining multijurisdictional collaboration. Both of these involved clusters of American states and Canadian provinces and once seemed promising models for climate policy collaboration between subfederal units in the two nations. However, one new initiative launched in November 2011 holds out some promise of renewed engagement of this type. The creation of "North America 2050: A Partnership for Progress" (NA 2050) was intended to facilitate provincial and state government efforts to establish and implement policies that reduce greenhouse gas emissions and create economic opportunities. This builds on the experience of the Three-Regions Collaborative, whereby leaders of the WCI, RGGI, and MGGRA launched efforts to share information and consider possible linkages between these proposed carbon trading systems. NA 2050 is open to all American and Mexican states and Canadian provinces and includes the goals of identifying "new leadership opportunities as climate and energy policy in North America continues to evolve."

It remains unclear just what this initiative will attempt to do, although it has proposed creation of working groups for policy analysis, the electricity sector, industrial energy efficiency, sequestration, biomass, offsets, and interprogram linkage. There are also additional signs that American states are exploring new ways to consider possible collaboration, despite the differences among some of them noted in earlier sections. The Environmental Council of the States (ECOS), for example, began a new initiative in 2011 to challenge states to work together to develop a "common GHG reduction policy." ECOS represents the lead environmental agency officials of the 50 states and has faced some rifts between climate leader and laggard states. But the new strategy is designed to try to bridge those differences by naming Delaware (widely seen as a national leader) and Indiana (widely seen as a national laggard) as coconveners of an effort to develop a consensus ECOS policy on climate change. These examples illustrate continued potential for states to work together, in part in response to the absence of a comprehensive federal role.

Of course, individual and collective state action may receive a further impetus from the emerging federal attempt to apply air quality standards to carbon emissions. The repeated indicators from Washington that the EPA wants not only to learn about state best practices on climate mitigation but also to reward states with more flexibility in meeting federal mandates could be a powerful source of expanded state engagement, potentially leading to a state race to the top to curry favor with federal authorities. As California Air Resources Board Director Mary Nichols noted in 2013,

> Having worked with EPA in the past and knowing their inclination to try to encourage action at the state level, we would expect them to bend over backwards to find ways to encourage and accommodate states that have already been moving . . . on pricing carbon.[27]

Whither Taxation and Budgets?

The 2012 national election was rapidly followed by intense battles over the federal budget and deficit as well as comparable debates in many states facing revenue shortages. Numerous proposals for far-reaching reforms of federal programs and the federal tax system emerged, including those from a series of commissions charged with developing a viable, long-term fiscal strategy for the nation. These have frequently included some reference to increased energy taxation or some form of a federal consumption tax as a primary way to raise additional revenue and thereby discourage consumption. It is impossible to anticipate the outcome of the coming debates, but it should be noted that it is possible that one or more versions of tax reform could serve to increase the levels of taxation applied to energy consumption, possibly moving toward the type of implicit carbon price noted earlier. This might parallel the experience abroad, such as that in the Canadian province of Ontario, which has harmonized its sales tax with the federal government and thereby increased taxation on energy as part of a larger tax base shift.

In turn, these discussions coincide with discussion of major questions surrounding the future of federal and state support for transportation infrastructure. Existing federal and state gasoline taxes produce declining yields given increased vehicle fuel economy, triggering considerable exploration for alternative revenue sources, most likely through some expanded levy on fuel or transportation or some new form of taxation, such as a charge per vehicle-mile traveled. Public opinion surveys consistently find that American support for expanded energy taxation increases significantly when the revenues are allocated toward popular programs, such as infrastructure development and repair or alternative energy development. Yet again, the United States may ultimately back into a suite of policies that collectively serve to reduce greenhouse gas emissions, even though their primary emphasis is on raising revenue.

Indeed, similar conversations are under way in many states, which are facing their own fiscal challenges and considering alternative revenue streams. Numerous states began to consider possible gas tax increases or reforms in 2013, all with the intent of generating a larger and steadier body of revenue to maintain transportation infrastructure. One possible model here might be just over the American border, where the Canadian province of British Columbia enacted a carbon tax in the late 2000s that reached $30 per ton in 2012 and returned all revenue to citizens via reductions in other taxes. This policy was proposed initially by a center-right government and has endured changes in political leadership, having demonstrated some capacity to reduce emissions, not deter economic growth, and sustain political support.

Suggested Readings

Craik, Neil, Debora VanNijnatten, and Isabel Studer, eds. *Climate Change Policy in North America: Designing Integration in a Regional System.* Toronto: University of Toronto Press, 2013.

Harrison, Kathryn, and Lisa Sundstrom, eds. *Global Commons, Domestic Decisions: The Comparative Politics of Climate Change.* Cambridge, MA: MIT Press, 2010.

Rabe, Barry G., ed. *Greenhouse Governance: Addressing Climate Change in the United States.* Washington, DC: Brookings Institution Press, 2010.

References

American Council for an Energy-Efficiency Economy (ACEEE). *The 2011 State Energy Efficiency Scorecard,* Research Report E115. Washington, DC: ACEEE, 2011.

Bianco, Nicholas M., and Franz T. Litz. "Old Roads to a New Destination." *Environmental Forum* 28, no. 3 (May/June 2011): 28–33.

Breggin, Linda. "Building Building Energy Codes." *Environmental Forum* 29, no. 4 (July/August 2011): 10.

Center for Local, State, and Urban Policy and Muhlenberg Institute of Public Opinion. *Public Opinion on Climate Policy Options.* Ann Arbor, MI: Center for Local, State, and Urban Policy, December 2012.

Cook, Brian J. "Arenas of Power in Climate Change Policymaking." *Policy Studies Journal* 38, no. 3 (2010): 465–486.

DiCosmo, Bridget. "McCarthy Pushes GHG Emission Cuts, Energy Efficiency as EPA Priorities." *Clean Energy Report,* InsideEPA.com, September 6, 2013.

Eisinger, Douglas S. *Smog Check: Science, Federalism, and the Politics of Clean Air.* Washington, DC: RFF Press, 2010.

Harrison, Kathryn. "Federalism and Climate Policy Innovation: A Critical Reassessment." *Canadian Public Policy* 39, supplement (2013): 95–109.

Harrison, Kathryn, and Lisa McIntosh Sundstrom, eds. *Global Commons, Domestic Decisions: The Comparative Politics of Climate Change.* Cambridge, MA: MIT Press, 2010.

Hoffmann, Matthew J. *Climate Governance at the Crossroads.* Toronto: University of Toronto Press, 2011.

Kahn, Debra. "Climate Chief Welcomes Federal Emissions Standard, Is Confident Cap and Trade Will Be Preserved." *E&E News,* June 26, 2013.

Knight, Chris. "EPA Asks States for Ideas on How to Cut Existing Power Plants." *Clean Energy Report,* InsideEPA.com, August 28, 2013.

Mann, Thomas E., and Norman J. Ornstein. *It's Even Worse Than It Looks.* New York: Basic Books, 2012.

Meckling, Jonas. *Carbon Coalitions: Business, Climate Politics, and the Rise of Emissions Trading.* Cambridge, MA: MIT Press, 2012.

Pooley, Eric. *The Climate War: True Believers, Power Brokers, and the Fight to Save the Earth.* New York: Hyperion Books, 2010.

Rabe, Barry G. "Contested Federalism and American Climate Policy." *Publius: The Journal of Federalism* 41, no. 3 (2011): 494–521.

Rabe, Barry G. *Statehouse and Greenhouse.* Washington, DC: Brookings Institution Press, 2004.

Rabe, Barry G. "States on Steroids: The Intergovernmental Odyssey of American Climate Policy." *Review of Policy Research* 25, no. 2 (2008): 105–128.

Rabe, Barry G., and Christopher P. Borick. "Carbon Pricing and Policy Labeling: Lessons From the American States and Canadian Provinces." *Review of Policy Research* 29, no. 2 (2012): 358–382.

Renewable Energy & Energy Efficiency Partnership (REEEP), Alliance to Save Energy, and American Council on Renewable Energy. *Compendium of Best Practices: Sharing Local and State Successes in Energy Efficiency and Renewable Energy from the United States.* Washington, DC, and Vienna, Austria: Renewable Energy & Energy Efficiency Partnership (REEEP), Alliance to Save Energy, and American Council on Renewable Energy, 2010. http://www.reeep.org/sites/default/files/Compendium%20of%20 US%20Best%20Practices.pdf.

Richardson, Nathan, Madeline Gottlieb, Alan Krupnick, and Hannah Wiseman. *The State of State Shale Gas Regulation.* Washington, DC: Resources for the Future, 2013.

Selin, Henrik, and Stacy Van Deveer, eds. *Changing Climates in North American Politics.* Cambridge, MA: MIT Press, 2009.

Sharp, Phil. "U.S. Energy Policy: A Changing Landscape." *Resources,* no. 181 (2012): 23–28.

Thomson, Vivian E., and Vicki Arroyo. "Upside-Down Cooperative Federalism: Climate Change Policymaking and the States." *Virginia Environmental Law Journal* 29 (2011): 1.

Wheeler, Steven. "State and Municipal Climate Plans: The First Generation." *Journal of the American Planning Association* 74, no. 4 (2008): 481–496.

Notes

1. Matthew J. Hoffmann, *Climate Governance at the Crossroads* (Toronto: University of Toronto Press, 2011); Henrik Selin and Stacy VanDeveer, eds., *Changing Climates in North American Politics* (Cambridge: MIT Press, 2009).
2. Barry G. Rabe, *Statehouse and Greenhouse* (Washington, DC: Brookings Institution Press, 2004); Rabe, "Contested Federalism and American Climate Policy," *Publius: The Journal of Federalism* 41, no. 3 (Summer 2011): 494–521.
3. Kathryn Harrison and Lisa McIntosh Sundstrom, eds., *Global Commons, Domestic Decisions: The Comparative Politics of Climate Change* (Cambridge, MA: MIT Press, 2010).
4. Thomas E. Mann and Norman J. Ornstein, *It's Even Worse Than It Looks* (New York: Basic Books, 2012).
5. Eric Pooley, *The Climate War: True Believers, Power Brokers, and the Fight to Save the Earth* (New York: Hyperion, 2010).
6. Jonas Meckling, *Carbon Coalitions: Business, Climate Politics, and the Rise of Emissions Trading* (Cambridge: MIT Press, 2012).
7. Nicholas M. Bianco and Franz T. Litz, "Old Roads to a New Destination," *Environmental Forum* (May-June 2011), p. 28.
8. Bridget DiCosmo, "McCarthy Pushes GHG Emission Cuts, Energy Efficiency as EPA Priorities," *Clean Energy Report,* InsideEPA.com (September 6, 2013); Chris Knight, "EPA Asks States for Ideas on How to Cut Existing Power Plants' GHGs," *Clean Energy Report,* InsideEPA.com (August 28, 2013).
9. Douglas S. Eisinger, *Smog Check: Science, Federalism, and the Politics of Clean Air* (Washington, DC: RFF Press, 2010).
10. Rabe, "Contested Federalism."
11. Steven Wheeler, "State and Municipal Climate Plans: The First Generation," *Journal of the American Planning Association* 74, no. 4 (2008): 481–496.
12. Eisinger, *Smog Check.*
13. Kathryn Harrison, "Federalism and Climate Policy Innovation: A Critical Reassessment," *Canadian Public Policy* 39, supplement 2 (2013): 95–109.
14. Barry G. Rabe, "States on Steroids: The Intergovernmental Odyssey of American Climate Policy," *Review of Policy Research* 25, no. 2 (March 2008): 38.
15. Renewable Energy & Energy Efficiency Partnership, *Compendium of Best Practices: Sharing Local and State Successes in Energy Efficiency and Renewable Energy from the United States* (Washington, DC: REEEP, 2010), 19.
16. American Council for an Energy-Efficiency Economy, *The 2011 State Energy Efficiency Scorecard* (Washington, DC: ACEEE, 2011), viii.
17. Linda Breggin, "Building Building Energy Codes," *Environmental Forum* 29, no. 4 (July/August 2012): 10.
18. Vivian E. Thomson and Vicki Arroyo, "Upside-Down Cooperative Federalism: Climate Change Policymaking and the States," *Virginia Environmental Law Journal* 29 (2011): 1.
19. Center for Local, State, and Urban Policy and Muhlenberg Institute of Public Opinion, *Public Opinion on Climate Policy Options* (Ann Arbor, MI: Center for Local, State, and Urban Policy, December 2012).

20. Neil Craik, Debora VanNijnatten, and Isabel Studer, eds., *Climate Change Policy in North America: Designing Integration in a Regional System* (Toronto: University of Toronto Press, 2013).

21. Brian J. Cook, "Arenas of Power in Climate Change Policymaking," *Policy Studies Journal* 38, no. 3 (2010): 465–486.

22. Barry G. Rabe and Christopher P. Borick, "Carbon Pricing and Policy Labeling: Lessons From the American States and Canadian Provinces," *Review of Policy Research* 29, no. 2 (May 2012): 358–382.

23. Phil Sharp, "U.S. Energy Policy: A Changing Landscape," *Resources* 181 (2012): 24.

24. Ibid.

25. Nathan Richardson, Madeline Gottlieb, Alan Krupnick, and Hannah Wiseman, *The State of State Shale Gas Regulation* (Washington, DC: Resources for the Future, 2013).

26. Center for Local, State, and Urban Policy and Muhlenberg Institute of Public Opinion, *Public Opinion on Climate Policy Options*.

27. Debra Kahn, "Climate Chief Welcomes Federal Emissions Standard, Is Confident Cap and Trade Will Be Preserved," *E&E News* (June 26, 2013).

CHAPTER 4

Climate Policy Innovation in American Cities

Rachel M. Krause

Introduction

Climate policy at the federal level in the United States has been characterized by conflict and stagnation. However, a considerable—and some say surprisingly large—amount of action to reduce greenhouse gas (GHG) emissions is being taken by subnational governments. In the previous chapter, Barry G. Rabe describes climate policies in US states, and this chapter examines the initiatives being pursued by municipalities. Over 1,000 US cities, accounting for approximately 30 percent of the country's total population, have pledged to reduce their GHG emissions.[1] These pledges and any follow-up actions taken to fulfill them are completely voluntary, since neither the federal government nor any state has passed legislation requiring municipal abatement. Even California's Senate Bill 375, which directs local governments to consider climate change in transportation and land use planning, does not mandate that cities actually achieve reductions. Individually, cities' efforts represent little more than a drop in the bucket in terms of the nation's net emissions; however, when taken together their actions have the potential to make a meaningful impact. This idea underlies much of the motivation for local initiatives.

Yet, it is exactly this notion of collective action that makes meaningful action on climate change difficult in general, not to mention for the smallest units of governments, where its effect is expected to be magnified.[2] Carbon dioxide dissipates globally, so the environmental benefits of its reduced atmospheric concentration are the same no matter where in the world emissions are reduced. Each abating entity receives little direct advantage from its abatement efforts, yet bears the full burden of the cost.[3] This dynamic creates an incentive to free ride and helps explain the dominant national and international focus of climate policy: It is assumed that in a noncoercive environment, neither governments nor individual emitters will choose to accept the extra costs of GHG abatement. Moreover, it is commonly thought that a conducive environment needs to be created in each country through policy implemented at the national level, which is best achieved through an international agreement. Thus, the fact that, in relatively large numbers, local governments are *voluntarily* choosing to engage in climate protection is somewhat of a surprise and, at least at first glance, appears to be an exception to the well-established theory of collective action.

Taking a somewhat different angle, Bai (2007) describes issues of scale and readiness as the largest obstacles to local climate participation. Specifically, she points to three common perceptions that inhibit municipal involvement: (1) It is beyond local governments' physical boundaries of concern; (2) it is beyond the time horizon relevant to urban political contexts where leaders, funding, and hot-button issues undergo frequent and rapid change; and (3) it simply is not local governments' business to address transboundary environmental concerns. Thus, from a variety of perspectives, local involvement, much less leadership, in climate policy is unexpected.[4]

More recently, however, the idea that voluntary local climate initiatives are indeed rational has begun to gain footing. This is due to a growing recognition of the positive side effects or cobenefits that often accompany them. For example, many GHG mitigation efforts involve increasing energy efficiency, which can translate directly into savings on electricity and fuel costs. Alternatively, innovative climate protection initiatives may result in positive attention being given to a city, potentially enhancing its image as a good place to live and invest. Organizations that promote local climate protection, such as ICLEI–Cities for Sustainability, highlight cobenefits as a way to help hook cities on the idea and use terms like *climate prosperity* to emphasize the point.[5] Thus, although the *climatic* benefit of GHG abatement is a public good, to which the theory of collective action applies, the cobenefits that accompany climate protection initiatives suggest that cities are at least partially motivated by the pursuit of local gains.

A three-part frame that characterizes the relationship between cities and climate change offers additional rationale for their involvement in its mitigation. According to this frame, cities are significant contributors to the problem of climate change, are expected to suffer disproportionately from it, and are strategically positioned to bring about reductions in GHG emissions.[6] An estimated 30–40 percent of global anthropogenic greenhouse gas emissions come from within cities' physical boundaries,[7] with significantly more being driven by urban demands but released elsewhere. Along with contributing to the problem, urban areas are particularly vulnerable to the effects of climate change. Cities are expected to experience above-average increases in temperature and other projected climate change impacts (e.g., increased floods, droughts, storms) likely to have a greater human and financial toll in cities because of their concentrated people and investments.[8] Finally, cities are well suited to be part of a climate solution: They have authority over a number of climate-relevant activities—often including land use and transportation planning, building codes, and waste disposal. They are also the level of government closest to the citizen and may be able to influence local resource use in a more targeted way than other levels of government. This characterization of a multipronged city–climate change relationship presents cities as having both the imperative and the ability to engage in its mitigation.

In addition to mitigation, some local governments are also pursuing policies aimed at facilitating adaptation to climate change. The nature of climate-induced threats varies by geography and often includes increased risk of floods, wildfires, drought, or extreme heat. For each relevant threat, adaptation requires cooperation from a different set of actors, with objectives that range from immediate disaster relief to long-term prevention planning. Adaptation, unlike mitigation, provides direct local benefits and thus should not be hindered by the barriers induced by collective action. Nonetheless, most cities have given far less attention to climate adaptation.

This chapter focuses on climate change mitigation initiatives that have been implemented in US cities and addresses three overarching questions: What are cities doing? Why are they doing it? and How much do their efforts matter? The next section provides a brief history of local climate protection in the United States, tracing it from its emergence in the early 1990s to the present. This is followed by a review of the different types of actions cities are implementing to reduce their GHG emissions. The next section describes the drivers and obstacles to local climate initiatives and presents the primary explanations for variation in involvement between cities. The final section discusses the value of local efforts given the global nature of climate change. A brief case study accompanies each of the last three sections and illustrates the concepts discussed in the context of a specific city.

A Brief History of Local Climate Policy

Local climate policy in the United States extends over about 20 years, and its evolution can be characterized as having three "eras of action." Although their start and end dates are fuzzy, the first era spans from about 1991 to 2004 and is a time in which a small number of cities acted as climate pioneers. The second, from about 2005 to 2010, witnessed a rapid acceleration in the spread of climate commitments across US cities; and the third (mid-2010 to the present) is characterized by a slowing in the number of new cities joining the movement and a reconsideration of efforts by previously committed cities. Looking back on the path this policy issue has taken gives insight into the factors that led to local action becoming a significant part of the climate protection movement and provides context to forecast its future trajectory.

Era 1: Climate Pioneers (1991–2004)

Intentional local involvement in global sustainability and climate issues is often traced back to the 1992 United Nations Conference on Environment and Development (often referred to as the Rio Earth Summit). The major output of this event was Agenda 21, a nonbinding but (at the time) widely embraced agreement that set out an action plan for addressing global sustainability problems in the 21st century. Chapter 28 of this document has become known as Local Agenda 21 and presents a rationale for local involvement in global issues. It is often considered an initial enabling force for cities' involvement in climate policy.

The first and arguably most important transnational network of local governments focused on increasing sustainability, the International Council for Local Environmental Initiatives (ICLEI), was an outgrowth of the UN Environment Programme. One of its first initiatives was the Urban C02 Reduction Project, which had the mission to "develop comprehensive local strategies to reduce greenhouse gas emissions and quantification methods to support such strategies."[9] Fourteen municipalities worldwide participated in this program, including six from the United States: Chula Vista, California; Dade County, Florida; Denver, Colorado; Minneapolis and St. Paul, Minnesota; and Portland, Oregon. The Urban C02 Reduction Project ran for two years before transitioning in 1993 to the Cities for Climate Protection (CCP) program, which remains a considerable player in local climate policy today.

Early local movers on climate change often had prior interests in environmental and quality-of-life issues and were able to link the policy of GHG reduction to an existing issue on the local agenda (e.g., poor air quality). They also tended to have the active support of a local official—often a mayor—who pushed for local action.[10] After committing to a climate agenda, many pioneer cities initially targeted their own municipal operations and low-hanging fruits, which were often efficiency efforts that could be justified by cost savings. Denver, Colorado, was a climate pioneer and exhibited many of these characteristics. Its mayor in the early 1990s, Wellington Webb, was a powerful force behind the city's participation in the Urban Reduction Project and ICLEI CCP. The city's climate involvement was linked to his larger vision of establishing Denver as an environmental leader and connected with previous local interests in renewable energy and reducing air pollution. Denver's initial follow-through actions were self-administered and included the creation of the nation's first municipal green fleets program and the installation of LEDs in red traffic lights and "don't walk" signs in the city's 1,200 intersections.[11]

Era 2: Spread of (Symbolic) Commitments (2005–2010)

In 2005 the Kyoto Protocol entered into force for the 141 countries that had signed and ratified it. This provided a jolt to what had been a relatively small and quiet movement of cities engaging in climate protection initiatives. As an act of protest over the United States' nonparticipation, Seattle Mayor Greg Nickels launched the Mayors' Climate Protection Agreement (MCPA). The MCPA provided a public forum for local leaders to express their support for climate protection and commit to local action. Cities that signed on to the agreement pledged to work to reduce their GHG emissions to 7 percent below their 1990 levels—the same amount as the United States would have been required to abate under the Kyoto Protocol—and to lobby their state governments and the federal government to enact comprehensive climate policy legislation. At its annual meeting in June 2005, the US Conference of Mayors unanimously approved the MCPA and set up a center to support its objectives. As shown in Figure 4.1, the number of MCPA city signatories increased almost seven-fold in the five-year period between 2005 and 2010, reflecting the momentum and excitement behind the idea that, through the leadership of local efforts, the United States could meaningfully engage in GHG reduction.

Although the pace at which new cities jumped on the climate protection bandwagon during this time was viewed positively by many proponents, it also

FIGURE 4.1   Number of Municipal Members in ICLEI CCP and MCPA Over Time

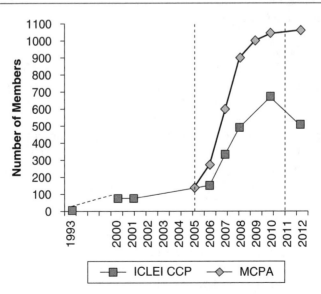

suggested that some were committing to objectives without having seriously considered the policies they would implement to achieve them. Some of these newly committed cities, such as New York City and Austin, developed comprehensive, high-profile plans to achieve GHG reductions. However, examinations of larger numbers of cities show uneven levels of effort and suggest that while some had indeed implemented meaningful local climate policy, other "committed" locations had done little more than engage in a symbolic gesture.[12] According to Betsill and Bulkeley, this has contributed to the recognition of a "stubborn gap between rhetoric and reality of local climate policy."[13]

Era 3: Reconsideration, Retrenchment, Recommitment (2010–)

The last several years have seen a decline in the momentum behind local climate protection initiatives. Whether because of increased public opposition to climate efforts, because of recent financial pressures facing local governments, or simply because the cities with the capacity to address this issue have already done so, the rate at which new cities are beginning explicitly to engage in climate policy has notably slowed, at least as measured though membership in the two primary networks of such cities.

Many previously committed local governments also appear to be engaging in reevaluations of their initiatives. In some cases this has led to withdrawal and in others, recommitment. This reevaluation is in part due to the increased external pressure cities are receiving from some segments of the public. Local climate efforts, particularly those that involve ties with ICLEI, have recently come under increased scrutiny. For example, Missoula, Montana, was a pioneer in the climate protection movement and has been an ICLEI member since 1996. After 15 years of relatively uncontroversial involvement, it experienced considerable opposition to its 2012 membership renewal. ICLEI, which originated as a United Nations initiative, has been singled out for opposition by right-wing groups, which view it as a threat to liberty and property rights. After a contentious two-hour public comment period, during which the police were called, the city council, in a split decision, voted to renew its ICLEI membership.[14] In the year since, Missoula has redoubled its climate commitment, and the city council unanimously passed a Conservation and Climate Action Plan that sets out the goals and strategies for the city to become carbon neutral by 2025.[15] In other locales, particularly in Tea Party strongholds like Texas, Florida, and Virginia, the anti-ICLEI movement has been more successful at getting cities to walk away from their climate protection initiatives.[16]

In sum, after more than a decade of involvement by a small minority of pioneer local governments, the decision to (at least rhetorically) engage in climate protection spread rapidly across US cities. More recently, the politics that has stymied federal climate legislation appears to have "trickled down" and, in some jurisdictions, the enthusiasm surrounding local climate policy has faded. Rather than being a widely embraced local policy issue, there is the potential that cities' explicit climate actions—and implementation of ideals of sustainability more broadly—will be increasingly determined by prevailing political ideologies. Still, the absolute number of strong climate-committed cities and their innovative nature solidify the leadership role of local governments in US climate policy.

What Are Cities Doing? The Content of Local Climate Policies

Many US cities have made public climate protection commitments, but what are they actually doing to reduce GHG emissions? There is significant variation in the strength, type, and comprehensiveness of cities' climate efforts. Some cities' initiatives are composed of ad hoc sets of seemingly unrelated actions, whereas others are more thoughtfully linked to GHG reduction goals. Two tools, GHG inventories and climate action plans (CAP), are considered key elements for serious local climate efforts.

A GHG inventory involves identifying and measuring all of the GHG emissions coming from a designated entity or physical area. Local inventories typically either target a city government's operation or a community at large. The importance of inventories is tied to the axiom, "If you can measure it, you can manage it." They enable cities to develop targeted policies and track emissions over time. The process of conducting GHG inventories is time and data intensive and requires technical expertise. ICLEI offers training and software to assist its members with this task, and a large portion of the US cities with completed inventories have utilized its assistance. While inventories can effectively measure change within a single city, they are less useful for cross-city comparisons because of differences in decisions made about scope. For example, some cities include emissions from local airport operations and from commuters who regularly drive, but do not live, in the city. Others use a narrower scope and do exclude these components. Protocols have recently been developed to try to limit variation in inventory structure across cities.[17] Many cities use their GHG inventories as a basis for identifying reduction goals and developing CAPs. (See Table 4.1 for the abatement goals adopted by the 20 largest cities in the United States.) CAPs detail the specific actions that cities intend to take to achieve specified reductions in their GHG emissions. CAPs are cognizant of local context and growth patterns and often build upon a GHG inventory. When well designed, CAPs include timelines, funding mechanisms, and the assignment of administrative responsibility to specified individuals or departments. As of late 2009, 56 cities had completed a CAP, and another 85 had one in progress.[18]

Cities can take a wide variety of actions as part of their efforts to reduce GHGs. The following two approaches help categorize them. The first differentiates

TABLE 4.1   GHG Emissions and Reduction Targets of the 20 Largest Cities in the
United States

City	Population	Reduction target
New York, NY	8,175,133	30% below 2005 levels by 2030
Los Angeles, CA	3,792,621	35% below 1990 levels by 2030
Chicago, IL	2,695,598	25% below 1990 levels by 2020
		80% below 1990 levels by 2050
Houston, TX	2,099,451	18% below 2010 business as usual projection for municipal operations
Philadelphia, PA	1,526,006	20% below 1990 levels by 2015
Phoenix, AZ	1,445,632	5% below 2005 levels by 2015
San Antonio, TX	1,327,407	7% below 1990 levels by 2012
San Diego, CA	1,307,402	Reduce regional emissions to 1990 levels by 2020
Dallas, TX	1,307,402	7% below 1990 levels by 2012
San Jose, CA	945,942	Reduce GHG emissions to 1990 levels by 2020
Jacksonville, FL	821,784	7% below 1990 levels by 2012
Indianapolis, IN	820,445	7% below 1990 levels by 2012
San Francisco, CA	805,235	25% below 1990 levels by 2017
		40% below 1990 levels by 2025
		80% below 1990 levels by 2050
Austin, TX	790,390	7% below 1990 levels by 2012
		Make all city operations carbon neutral by 2020
Columbus, OH	787,033	2% per year until 2030 (40% below 2005 levels)
Fort Worth, TX	741,206	7% below 1990 levels by 2012
Charlotte, NC	731,424	7% below 1990 levels by 2012
Detroit, MI	713,777	7% below 1990 levels by 2012
El Paso, TX	649,121	7% below 1990 levels by 2012
Memphis, TN	646,889	None

between actions that focus on abating emissions from municipal government operations and those that focus on the broader community.[19] The second differentiates between explicit climate actions, which are solely focused on climate protection, and actions that result in other benefits along with GHG reduction. This latter set of actions is sometimes referred to as "implicit climate protection."[20] Together, they enable the construction of a 2x2 grid characterizing relevant actions that local governments can undertake. Table 4.2 presents this grid and offers examples of local actions falling within each quadrant.

Government- and Community-Focused Initiatives

Many local governments engaging in climate protection differentiate between the actions they take that are focused on their own operations and those that target the entire community.

The former are more frequent and often appear to be how cities initiate their climate protection efforts. Very few cities develop community-orientated CAPs without first having one for their own government operations.[21] Bae and Feiock examine 13 energy- and climate-related issues and compare the number of cities that have addressed each for government facilities and the community at large.[22] Based on the survey responses of 679 cities, they find that, with only one exception (the provision of alternative transportation), policies that focus internally toward government operations are more common than comparable ones that target the broader community. In fact, for the following six energy and climate issue areas, the implementation gap exceeds 25 percent: building retrofitting (68 government/36 community); green procurement (51/8), technology innovation/demonstration projects

TABLE 4.2   Characterization and Examples of Local Climate Protection Activities

	Government operation focused	*Community focused*
Explicit climate objective	1. Inventoried GHG emission from government operations	1. Inventoried GHG emissions from government operations
	2. Developed climate action plan for government operations	2. Developed climate action plan for government operations
	3. Established an interdepartmental climate committee	3. Instituted outreach/ education efforts on climate change mitigation
Cobenefit generating actions	1. Established green building standards for city buildings	1. Established citywide green building code
	2. Changed city traffic lights to LED	2. Provided public transportation
	3. Transitioned to a green municipal vehicle fleet	3. Provided curbside recycling to city residents

(49/21); alternative fuels (51/20); energy efficient devices (e.g., appliances, lighting) (69/40); energy efficient systems (e.g., building controls) (69/28).

The difference in implementation rate is notable in part because, although they have been receiving the bulk of the attention, local government operations account for only a small portion of a community's total emissions—on average, under 3 percent.[23] However, although the direct impact of government-focused initiatives is limited in size, a high degree of control and certainty exists over their outcome. Through the self-administration of ordinances and internal rules, leaders are able to directly change city governments' GHG-relevant operations. Many of these efforts also result in cobenefits that accrue directly to the city, such as energy savings. Community emissions, on the other hand, compose the majority of total city emissions, so actions taken to influence them have the potential to be much more significant. However, most policies aimed at altering community emissions (i.e., public behavior) take the form of information, service provision, or incentives, so the control over their outcome is limited. Their implementation also tends to cost the city government resources (e.g., money, political capital) as opposed to creating direct benefits.

Explicit and Cobenefit Generating (Implicit) Actions

Explicit climate actions are clearly tied to a climate protection objective and result in the sole benefit of facilitating GHG reduction. Implicit climate actions result in multiple simultaneous benefits, only one of which is GHG abatement. As such, implicit actions may be motivated by reasons other than climate protection. If a climate-protecting intent is necessary for local actions to be considered part of climate policy, the treatment of implicit actions becomes challenging and could lead to the same local action being treated as a component of climate policy in one city and not in another. Consider the question of whether a city's public transportation services should be considered part of a climate protection initiative. Is the answer "yes" only if the city frames them as such? Is it "yes" only if they were initiated after a climate commitment was made? Is it "yes" only if it is included in a CAP?

Recognizing this complication, actions that produce benefits other than GHG reduction are not considered explicit. Using this distinction, Table 4.3 shows the implementation rates for 26 climate-protecting actions that are within the jurisdiction of local governments.[24] The table is divided into two sections with the top portion containing those actions that that have the single explicit goal of advancing GHG reduction and the bottom containing those that simultaneously contribute to GHG reduction and other local goals. The explicit actions are all either planning or administration oriented. A larger number are implicit, leading to the conclusion that a majority of things local governments can do to reduce GHGs result in other local benefits as well. The frequency of implementation ranges widely across activities:

from 17 to 56 percent for implicit actions and from 22 to 91 percent for explicit ones. Perhaps not surprisingly, cities have implemented cobenefit producing implicit actions more frequently than explicit ones. Moreover, the relatively small portion of cities (15 percent) that have developed and adopted a CAP suggests that most cities are taking relatively ad hoc project-based approaches to GHG reduction.

Table 4.3   Percentage of Cities That Have Implemented GHG-Reducing Actions

Explicit Climate Protection Actions

1.	Citywide GHG emissions inventory completed	41
2.	GHG reduction goal formally adopted	30
3.	Comprehensive plan to achieve GHG reduction developed/adopted	15
4.	Responsibility for managing city's climate protection activities designated to city employee, department, or volunteer committee	56
5.	Funding for climate protection designated in city budget	17

Implicit/Cobenefit-Producing Climate-Protecting Actions

6.	Efficient lighting installed in city buildings	59
7.	Efficient lighting installed in city streetlights	33
8.	Energy Star–only purchase policy for city equipment and appliances adopted	31
9.	Efficiency standards (such as LEED) adopted for all new and retrofit city buildings	40
10.	Policy in place limiting the amount of time city-owned vehicles may idle	48
11.	Conversion to green city vehicle fleet underway	81
12.	Incentive programs in place encouraging city employees to travel to work using means other than a single-occupancy vehicle	32
13.	City purchases and/or produces alternative energy to power its own operations	37
14.	City provides information about how to increase energy efficiency to its residents	77
15.	City provides financial incentives to the public and/or developers to encourage energy-efficient new construction or improvements to existing buildings	41
16.	City requires efficiency standards (such as LEED) be met in new commercial and/or residential construction	22

17. Outreach and education provided to residents regarding privately owned trees	56
18. Municipal ordinances in place that dictate tree planting and/or removal specifications for developers	75
19. Public transport services provided to city residents	61
20. Incentives offered for residents to take public transit (e.g., free days, reduced fares)	25
21. Bicycle lanes exist in roadways	70
22. Communitywide hiking and cycling trails in place	64
23. Planning and zoning decisions involve explicit consideration of their impact on GHG emissions and/or sprawl	66
24. Separated yard waste composted or mulched	63
25. Curbside recycling provided to city residents	91

Source: These results are based on the responses to a 2010 survey that was sent to all 665 US cities with populations over 50,000. Usable responses were obtained from 329 cities, for a response rate of 49.5 percent (see Krause, "Policy Innovation, Intergovernmental Relations," 2011).

A Closer Look: Kansas City, Missouri

Although it is unlikely to be described as a pioneer in the local climate protection movement, Kansas City, Missouri (KCMO), has taken substantial steps to mitigate its greenhouse gas emissions. For the purpose of illustrating what cities are doing in regard to climate change mitigation, it has the additional benefit of following a rather typical trajectory among active US cities.

Climate protection in KCMO officially began in 2005 with then-Mayor Kay Barnes's signing of the Mayor's Climate Protection Agreement. City council passed an ordinance shortly thereafter authorizing the mayor to initiate a climate-planning process. City staff, with the assistance of ICLEI, completed baseline GHG inventories for the years 2000 and 2005. A steering committee of community leaders reviewed those inventories and made recommendations for action to be included in the city's climate action plan (CAP). Steering committee members, who included prominent representatives from the local utility, the metropolitan planning organization, the chamber of commerce, and several nonprofit organizations, were selected because they represented stakeholder groups whose support would be necessary for a plan's successful implementation and because their recommendations would carry weight with city council.[25] The climate planning process proceeded in two distinct phases. As many cities do, KCMO initially focused on its government operations, guided by the rationale that city leaders have a high degree of control over municipal emissions and that government should lead by example. The Phase 1 abatement goal called for the reduction of GHG emissions from city government operations by 30 percent below year 2000 levels by 2020. This goal, along with 32 actions specified to achieve it, was approved in early 2007. The following year, as part of the second phase, the same goal was adopted for the community at large along with an "aspirational goal" of achieving 80 percent below 2000 levels by 2050 and eventual carbon neutrality.[26]

The CAP, adopted in 2008, culminated the planning efforts made in both phases. It contains 55 action items falling under the five general categories of energy, carbon offsets and waste management, transportation, buildings and infrastructure, and policy and outreach. The specific actions it includes were determined using a "triple bottom line" approach in an attempt to ensure that, along with reducing GHGs, they would also maximize economic and social benefits. Federal funds from the 2009 American Reinvestment and Recovery Act enabled the city to implement its plan. KCMO's climate planning process has been credited with influencing other regional initiatives, including the signing of the Mayor's Climate Protection Agreement by over 20 mayors in the Kansas City metro region and the development of a Climate Protection Partnership by the Greater Kansas City Chamber of Commerce.[27]

KCMO used a sound process in planning what the city would do to reduce its GHG emissions. It followed the standard recommended route for climate active cities: getting commitment from city leaders, conducting a GHG inventory, setting a reduction goal, and developing a local CAP with the assistance of community stakeholders. The result was a comprehensive set of actions that emphasized the triple bottom line and had the buy-in of key groups in the community. That said, progress with the implementation of the plan is less clear. Whereas the documents and reports leading up to the adoption of the CAP are readily and publicly available, postadoption progress reports and evaluations are not.

Why Are Cities Engaging in Climate Protection? An Assessment of Motivations

More than 1,000 cities in the United States have pledged action to help mitigate climate change. This is a significant number, but it also means that well over 25,000 US cities have not. What factors influence whether cities make this commitment? What are the main motivations behind the decisions of those that have?

Figure 4.2 presents an overview of the drivers and obstacles that have been shown to influence the likelihood that cities will engage in climate protection. It presents three overarching categories of explanations. The first set is related to the larger interest group theory of political decision making. Several studies have found that higher numbers of environmental nonprofits in a city or demographics that correlate with environmental concern increase the likelihood that city governments are involved in climate protection. The strength of the manufacturing sector, which is often considered an organized interest that would oppose climate initiatives, does not appear to have any notable effect. Explanations based around local civic capacity and political leaning are also related to the interest group theory in that they view government action as a response to community pressure. *Civic capacity* represents the degree of participation and social capital present in a community and may be related to the strength of environmental interests. Findings suggest that cities with higher overall levels of civic capacity (whether environmentally orientated or not) are more likely to have sustainability policies in place. The political leaning of city residents has a somewhat less consistent effect. Climate change has been characterized by partisanship in the United States, and it has been hypothesized that cities with a greater share of residents who support the Democratic Party would be more likely to engage in climate protection than those that lean Republican. Interestingly, this assumption appears to hold only when considering explicit climate actions and not for the implementation of implicit GHG-reducing actions.[28] This inconsistency is represented by the dashed arrow in Figure 4.2.

FIGURE 4.2   An Overview of the Factors Influencing the Likelihood That Cities Are Engaged in Climate Protection

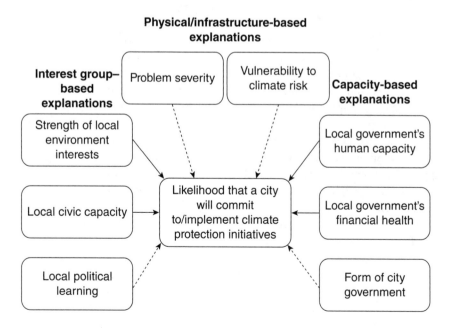

The next set of hypotheses points to cities' physical and infrastructure characteristics as factors that influence whether they decide to engage in climate protection. The first, problem severity, is based on the premise that GHG-emitting activities are more deeply embedded into the local physical and economic structures of some communities than others. Greater problem severity is thought to reduce the likelihood that cities will engage, because abatement would incur higher costs. The second, vulnerability to climate induced risk, recognizes that the location of some cities, like being on a coast, markedly increases the likelihood of their incurring damage as a result of climate change. Cities that are more vulnerable to climate change may be more likely to try to reduce it. Although both of these physical explanations seem logical, they have received uneven empirical support and appear to be secondary explanations at best.[29]

The third set of explanations for why some cities engage in explicit climate action and others do not focuses on characteristics of the local government itself. City governments that have high levels of capacity are thought to be more likely to engage in climate protection than those without it, all else equal. Fiscal health and human resources play a large role in determining governments' overall capacity. Strong fiscal conditions allow municipalities to adequately fund their programs and services as well as support more innovative priorities. Human resource capacity refers to both the number and ability of city employees. Previous studies reinforce the importance of both types of capacity and particularly point to the importance of skilled staff in climate involvement.[30] Larger cities tend to exhibit greater climate involvement, perhaps because population size is correlated with government capacity.

Finally, previous research shows that local institutional form can meaningfully affect which local interests get heard and the outcome of political decisions.[31] The relevant question in this context is whether having a mayor–council form of government—in which the head administrator (i.e., the mayor) is elected, or a council–manager form—in which the head administrator (i.e., the city manager) is appointed, influences the likelihood that a city will pursue climate protection. Several empirical studies have offered conflicting findings about the impact that government form has on climate engagement.[32] Bae and Feiock shed light on this inconsistency by separating local climate actions that target city government operations from those that target the community at large. They find that, all else equal, having a council–manager government results in more climate actions being implemented that focus on city government operations but have a negative effect on the number targeting the broader community.[33]

Figure 4.2 presents the factors that help explain why some cities become involved in climate protection and others do not. Figure 4.3, on the other hand, considers only the subset that have explicitly committed to it. It presents cities' characterizations of 11 potential motivations as extremely, somewhat, or not important factors in their decision to pursue climate protection as well as the single most important factor behind their decisions.

FIGURE 4.3   The Importance of Different Motivations Behind Cities' Explicit Climate
Protection Commitments

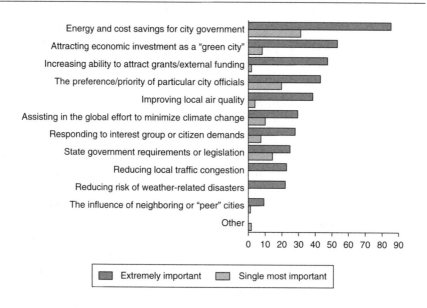

All of the considerations, with the exception of assisting in the global effort to minimize climate change, yield some form of local cobenefit, whether physical, economic, or political in nature. Figure 4.3 suggests that, for most cities, climate protection is itself the cobenefit rather than the primary objective. Only10 percent of cities describe assisting in global climate protection as the most important reason that they engage in GHG mitigation efforts, and a full 70 percent describe it as only a somewhat consideration or not an important consideration. Contributing to the public good of reduced global climate change appears, at best, to be a secondary motivation for many of the cities that are engaged in this issue.

By a large margin, city governments point to the desire to reduce energy-related expenses as their primary motivation. A full 85 percent describe it as an "extremely important" consideration, and nearly a third identify it as their single most important motivation. Economically orientated motivations hold other top spots, as does the need to accommodate the preferences and priorities of local government officials. The latter adds support to the observed importance of policy entrepreneurs in subnational climate policy.[34] Civic engagement and environmentally oriented interest group pressure appears to play a relatively modest role in most climate-engaged cities' decision to commit.

In sum, capacity—both in terms of the community's civic capacity and the financial and human capacity of the local government—and the strength of proenvironment interests offer the most consistent explanations for why some cities choose to pursue explicit climate protection initiatives and others do not. When asked why they decided to engage, committed cities overwhelmingly emphasize the economic motivation for their decisions. Capacity and a supportive public appear to be important factors enabling cities to take advantage of these cobenefits. The altruistic desire to assist in climate change mitigation does not play a very large role in most cities decisions.

A Closer Look: Chicago, Illinois

The City of Chicago adopted its climate action plan (CCAP) in 2008. CCAP proposes two GHG abatement goals—an initial goal of reducing emissions to 25% below 1990 levels by 2020 and an eventual goal of a 50% reduction by 2050—and describes the steps the city will take to meet them. The plan is developed around five main strategies: increasing energy efficiency in buildings, increasing the production and use of clean and renewable energy, improving transportation, reducing waste and industrial pollution, and preparing for climate adaptation. The CCAP strategies are based off of and monitored by a series of GHG inventories, which have been conducted for the years 2000, 2005, and 2010. An initial progress report details the actions implemented in the first two years after CCAP's adoption, which include 456 specific initiatives developed by city departments and agencies to help the city mitigate and adapt to climate change.[35]

Chicago is clearly a leader in local climate policy, and the reasons behind its decision and ability to successfully pursue this role illustrate many of the points made more generally in this section, specifically the importance of political leadership, human and financial resources, community support, and cobenefits. Longtime Chicago mayor Richard Daley provided the initial force behind the city's CCAP. He linked climate protection with a goal expressed early in his administration to develop a reputation for Chicago as "the most environmentally friendly city in the nation."[36] The strong influence of a powerful political entrepreneur who acted as a champion for the plan made it easier to bring together city departments and community leaders and secure necessary resources for the project.

The city also benefitted from having a high-capacity and sufficiently large staff in its Department of Environment (DOE), which was tasked with the CCAP's overall project management. The commissioner of the DOE, Sadu Johnson, was a high-level proponent of the initiative within the city bureaucracy (i.e., a policy entrepreneur) and has been credited with providing the processes with vision and leadership. In addition to human resources, the CCAP also had designated financial resources, much of which came from philanthropic and community organizations. Indeed Chicago's climate protection efforts benefited from consistent community support from a variety of local interests, including unions, the business community, universities, and the nonprofit sector. The role of many of these groups was formalized in the climate planning process by having representatives serve on a task force that advises city action. A close partnership with the nonprofit Global Philanthropy Partnership helped the city administer the processes and maneuver efficiently through technical and strategic issues that may have otherwise caused delay.[37]

Finally, cobenefits were an important part of the motivation behind Chicago's development of its CCAP. They receive considerable emphasis in publicly released communications about the effort and make it clear that each strategy solves multiple problems. Table 4.4 is from a CCAP evaluation document and shows how the city thinks about the benefits of its major strategies. Its emphasis on cost savings, jobs, and economic development reinforces the importance of economic cobenefits as motivations for climate engagement and may help justify the plan to members of the public who may be skeptical. In sum, Chicago's decision to pursue climate protection was primarily a result of mayoral initiative, the desire for a national green reputation, and the pursuit of economic cobenefits. Community support and a high-capacity staff facilitated its success and enabled it to become a climate leader.

TABLE 4.4   Cobenefits of Climate Policy for the City of Chicago

	Cobenefits						
Strategy	Reduced energy costs	Jobs	Economic development	Improved air quality and health	Water quality	More appealing communities	A more resilient city
Energy efficient buildings	✓	✓	✓	✓	✓	✓	✓
Clean & renewable energy sources		✓	✓	✓			✓
Improved transportation options	✓	✓	✓	✓		✓	✓
Reduced waste & industry pollution	✓	✓	✓	✓	✓	✓	✓
Adaptation to climate change	✓	✓	✓	✓	✓	✓	✓

Source: Parzen (2009).

What Impact Can Locally Led Initiatives Have?

In 2011, the United States emitted 6,702 million metric tons (mmt) of carbon dioxide equivalent (CO_2e) gases.[38] Given this quantity, can local efforts have any meaningful impact? Some experts claim that local action has an important role in any comprehensive climate strategy,[39] whereas others suggest that the amount of abatement that localities are capable of achieving is too small to have a meaningful impact on global climate change.[40] Arguments on both sides, however, have largely been based on logic and anecdotal evidence rather than empirical findings.

In one of the few empirical pieces on this topic, Lutsey and Sperling (2008) estimate that if all the US cities that had signed on to the Mayors' Climate Protection Agreement were to achieve their goals, by 2020 they would cumulatively abate approximately 597 million metric tons (mmt) of CO_2e, or slightly less than 9 percent of US annual emissions.[41] As described previously, many cities focus on reducing emissions from their own governmental operations. If cities successfully achieved carbon neutrality for their own government operations, as the City of Austin, Texas, has pledged to do, it would result in an estimated total annual reduction of about 136.7 mmt of CO_2e.[42] This is equivalent to the annual emissions generated by 37 coal-fired power plants and is about 2 percent of annual US emissions.[43] Both of these estimates rest on numerous assumptions and focus on potential accomplishments rather than actual ones. Efforts to quantify the impact of actual climate protection initiatives are limited by a lack of data, which prevents defensible assertions about the cumulative impact of local efforts from being made. That said, several individual cities—New York City, Chicago, and San Francisco, to name a few—have done a commendable job measuring and recording their GHG emissions over time. New York's efforts are examined in more detail in the next section.

In sum, the cumulative impact of local climate policies cannot be assessed with any certainty, but the amount of direct abatement they generate is probably small. That said, they may also result in meaningful indirect impacts. For example, municipal governments often "lead by example." When cities take visible climate-protecting actions, it sends the message that GHG reduction is a worthy goal and, moreover, is one that can be pursued within the local community. A local government's decision to install solar panels on the roof of its city hall could, for example, influence local businesses to do the same. Thus, its impact is not limited to the emissions abated directly through the city's own renewable energy use but also includes reductions resulting indirectly from its leadership. "Leading by example" is an explicit part of the rationale behind some municipal leaders' engagement in climate protection.[44]

A second potential indirect impact of local climate policies is based on purchasing power. Governments often intentionally structure their procurement policies to push the market in a proenvironment direction and spur innovation. The federal government and numerous state and local governments have self-imposed green procurement policies. Purchase commitments by cities for renewable energy and energy efficient goods can provide producers some certainty about demand, expand the market, and assist in the technology becoming economically competitive. Although it cannot be measured, the impact of local climate policy likely extends well beyond the actions and emissions it directly targets.

Finally, when local governments pursue climate action, they bring what can often feel like an abstract problem closer to home for members of the community. When climate-planning processes engage the citizenry, they educate citizens and increase public knowledge about the issue and can stimulate participants to look for ways to reduce their own carbon footprints. High-quality public participation has also been found to increase citizens' trust in public agencies and result in better assessments of their responsiveness and value as well as to increase tolerance for differences of opinion more generally.[45] Thus, although it cannot be measured, the impact of local climate policy often extends well beyond the actions and emissions it directly targets.

A Closer Look: New York City

A handful of high-capacity cities regularly monitor their GHG emissions and can provide some insight into questions about impact. Among these cities, New York City stands out. Its climate protection program is part of its larger PlaNYC initiative, which was launched in 2007 at the behest of Mayor Michael Bloomberg. It is a citywide effort that aims to strengthen the economy, prepare for climate change, and improve the quality of life for New Yorkers.[46] Since its establishment, six GHG inventories—one each year—have been released, resulting in what is arguably the most comprehensive and sophisticated set of municipal GHG inventories in the country.

Per capita emissions in NYC are far lower than the US average and have a considerably different composition. The operation of buildings creates a far larger percentage (74 percent), and transportation a far lower percentage (21 percent), of emissions than in most cities and the United States as a whole. Having a clear picture of its specific emission sources enables the city to target policies in a way that would have been impossible if national averages were relied on.

In the first seven years after the initial implementation of PlaNYC in 2005, net emissions decreased by 16.1 percent, despite an increase in the number of people and buildings in the city. Some of the reduction is attributed to external factors like mild weather and national vehicle fuel economy standards, but the larger portion is a result of reduced consumption and changes made to utility operations. Together, changes made in these two areas resulted in a reduction of 11.2 mmt CO_2e, which more than offset the increase caused by population growth.

The use of less carbon-intensive methods of electricity generation account for the single largest reduction: 7.01 mmt CO_2e or 11 percent of the city's net emission reduction since 2005.[47] This reinforces previous observations that having influence over utilities—particularly electric utilities—can greatly increase the impact of local policies. Over the past seven years, local policies in NYC resulted in annual GHG emissions being 11.2 mmt less than they otherwise would have been. This annual reduction is equivalent to taking 2.2 million cars of the road or shutting down three coal-fired power plants for a year. Put in these terms, the actual impact of one city's climate initiative, albeit a large city and a comprehensive climate policy, appears quite significant.

Conclusion

The past decade has seen unexpected and unprecedented levels of involvement by cities in the global issue of climate change mitigation. In terms of absolute numbers, local efforts account for a majority of the explicit climate policies adopted in the United States. However, because of the number, variation, and voluntary nature of local efforts, data about them are sparse, and there is relatively little generalizable knowledge about the intensity or outcome of their implementation.

The information that is comparable across cities includes whether they have expressed a commitment to climate protection as well as surface-level knowledge about the types of relevant actions they have implemented. Studies based on this information have revealed some general insights, including that municipal climate protection efforts appear to be heavily motivated by the cobenefits that often accompany them and that a high-capacity government and a supportive and engaged public are key factors that enable cities to take the necessary steps to pursue such efforts. That said, there is more to learn about the dynamics that produce the political and planning contexts that facilitate meaningful local action.

Urban policy making offers opportunity for direct citizen involvement. In part because of the proximity between constituent and decision maker, local governments are often more responsive to public demands than other levels of government. In some locales, like Seattle, citizen groups played a large role in initiating and shaping climate efforts. Case study based research by Portney and Berry also finds that cities with a more participatory public are more likely to have engaged in sustainability efforts.[48] It is difficult, however, to draw generalizable conclusions about the impact that civic engagement has on cities' climate efforts beyond that it tends to serve a positive enabling role. It does not appear to be primary: When asked directly, city staff at over 70 percent of US cities with an explicit climate protection effort described responding to interest group or citizen demands as only a "somewhat" or "not important" motivation for their climate initiatives.[49]

It is reasonable to expect that public engagement would play a larger role in initiating and shaping GHG reduction efforts that focus on the community at large, as opposed to those that focus on city government operations. Still, cities more frequently implement inwardly focused climate initiatives, even though the possible magnitude of their effect is much smaller. This may be because they often result in cost savings for the government and take place in a controlled arena, which makes them easier to carry out.

Although the role played by members of the community and local NGOs is not clear or consistent across cities, several climate networks have done much to draw attention to the contributions municipal governments can make to climate protection. ICLEI–Cities for Sustainability, the C40 Cities Climate Leadership group, the US Conference of Mayors' Climate Protection center, and the Sierra Club's Cool Cities program are among the more known of these networks. These organizations require different degrees of formal commitment from and offer varying levels of technical support to their member-cities. It is not clear if or how much participation in each of these networks increases the GHG-abating actions being implemented by member cities. However, these actions have successfully changed the overall issue environment and put climate change onto the urban policy agenda.

Suggested Readings

Bulkeley, Harriet, and Michele Betsill. *Cities and Climate Change: Urban Sustainability and Global Environmental Governance.* London and New York: Routledge, 2003.

Feiock, Richard C., and Christopher Coutts. "Climate Change and City Hall." Special Issue, *Cityscape* 15, no. 1 (2013). http://www.huduser.org/portal/periodicals/cityscpe/v0115num1/index.html.

ICLEI USA Annual Reports. Available from the ICLEI website: http://www.icleiusa.org/about-iclei/annual-reports.

PlaNYC 2030 Reports. Available from New York City's municipal website: http://www.nyc.gov/html/planyc2030/html/publications/publications.shtml.

Portney, Kent. *Taking Sustainable Cities Seriously: Economic Development, the Environment, and Quality of Life in American Cities,* 2nd ed. Cambridge, MA: MIT Press, 2013.

References

Aall, Carlo, Kyrre Groven, and Gard Lindseth. "The Scope of Action for Local Climate Policy: The Case of Norway." *Global Environmental Politics* 7, no. 2 (2007): 83–101.

Bae, Jungah, and Richard Feiock. "Forms of Government and Climate Change Policies in US Cities." *Urban Studies* 50, no. 4 (2013): 776–788.

Bai, Xuemei. "Integrating Global Environmental Concerns Into Urban Management: The Scale and Readiness Arguments." *Journal of Industrial Ecology* 11, no. 2 (2007): 15–29.

Betsill, Michele. "Mitigating Climate Change in US Cities: Opportunities and Obstacles." *Local Environment* 6, no. 4 (2001): 393–406.

Betsill, Michele, and Harriet Bulkeley. "Looking Back and Thinking Ahead: A Decade of Cities and Climate Change Research." *Local Environment* 12, no. 5 (2007): 447–456.

Bulkeley, Harriet, and Michele Betsill. *Cities and Climate Change: Urban Sustainability and Global Environmental Governance.* New York and London: Routledge, 2003.

City of New York. *PlaNYC: An Inventory of New York City Greenhouse Gas Emissions.* 2012. http://nytelecom.vo.llnwd.net/015/agencies/planyc2030/pdf/greenhousegas_2012.pdf.

Hunt, Alistair, and Paul Watkiss. "Climate Change Impacts and Adaptation in Cities: A Review of the Literature." *Climatic Change* 104, no. 1 (2011): 13–49.

International Council for Local Environmental Initiatives (ICLEI). *Local Government Implementation of Climate Protection: Report to the United Nations.* Toronto: International Council for Local Environmental Initiatives, 1997.

International Council for Local Environmental Initiatives (ICLEI) USA. *ICLEI's Five Milestones for Climate Mitigation.* http://www.icleiusa.org/climate_and_energy/climate_mitigation_guidance/iclei2019s-five-milestones-for-climate-mitigation.

Krause, Rachel M. "An Assessment of the Greenhouse Gas Reducing Activities Being Implemented in US Cities." *Local Environment* 16, no. 2 (2011): 193–211.

Krause, Rachel. M. "The Motivations Behind Municipal Climate Engagement: An Empirical Assessment of How Local Objectives Shape the Production of a Public Good." *Cityscape: A Journal of Policy Development and Research* 15, no. 1 (2013): 125–141.

Krause, Rachel M. "Policy Innovation, Intergovernmental Relations, and the Adoption of Climate Protection Initiatives by US Cities." *Journal of Urban Affairs* 33, no. 1 (2011): 45–60.

Krause, Rachel M. "Political Decision-Making and the Local Provision of Public Goods: The Case of Municipal Climate Protection in the US." *Urban Studies* 49, no. 11 (2012): 2399–2417.

Lutsey, Nicholas, and Daniel Sperling. "America's Bottom-Up Climate Change Mitigation Policy." *Energy Policy* 36, no. 2 (2008): 673–685.

Parzen, Julia. *Lessons Learned: Creating the Chicago Climate Action Plan.* Chicago: City of Chicago Department of Environment, 2009. http://www.chicagoclimateaction.org/filebin/pdf/LessonsLearned.pdf.

Peterson, D., E. Matthews, and M. Weingarden. *Local Energy Plans in Practice: Case Studies of Austin and Denver.* Golden, CO: National Renewable Energy Laboratory, 2011. http://www.nrel.gov/tech_deployment/state_local_activities/pdfs/50498.pdf.

Portney, Kent E., and Jeffrey M. Berry. "Participation and the Pursuit of Sustainability in US Cities." *Urban Affairs Review* 46, no. 1 (2010): 119–139.

Rabe, Barry G. *Statehouse and Greenhouse: The Emerging Politics of American Climate Change Policy.* Washington, DC: Brookings Institution Press, 2004.

Satterthwaite, David. "Cities' Contribution to Global Warming: Notes on the Allocation of Greenhouse Gas Emissions." *Environment and Urbanization* 20, no. 2 (2008): 539–549.

Selin, Henrik, and Stacy D. VanDeveer. "Political Science and Prediction: What's Next for US Climate Change Policy?" *Review of Policy Research* 24, no. 1 (2007): 1–27.

Sharp, Elaine B., Dorothy M. Daley, and Michael S. Lynch. "Understanding Local Adoption and Implementation of Climate Change Mitigation Policy." *Urban Affairs Review* 47, no. 3 (2011): 433–457.

US Environmental Protection Agency. *Inventory of US Greenhouse Gas Emissions and Sinks: 1990–2011,* EPA 430-R-13–001. Washington, DC: Environmental Protection Agency, 2013.

Wheeler, Stephen M. "State and Municipal Climate Change Plans: The First Generation." *Journal of the American Planning Association* 74, no. 4 (2008): 481–496.

Wiener, Jonathan B. "Think Globally, Act Globally: The Limits of Local Climate Policies." *University of Pennsylvania Law Review* 155, no. 6 (2007): 1961–1979.

Zahran, Sammy, Himanshu Grover, Samuel D. Brody, and Arnold Vedlitz. "Risk, Stress, and Capacity Explaining Metropolitan Commitment to Climate Protection." *Urban Affairs Review* 43, no. 4 (2008): 447–474.

Notes

1. Mayors' Climate Protection Center, List of Participating Mayors, http://www.usmayors.org/climateprotection/list.asp.
2. Yael Wolinsky-Nahmias, "Introduction: Global Climate Politics," in *Global Climate Politics,* edited by Yael Wolinsky-Nahmias (Thousand Oaks, CA: CQ Press, 2015), 1–30.
3. Mancur Olson, *The Logic of Collective Action: Public Goods and the Theory of Groups* (Cambridge, MA: Harvard University Press, 1965).
4. Xuemei Bai, "Integrating Global Environmental Concerns Into Urban Management: The Scale and Readiness Arguments," *Journal of Industrial Ecology* 11, no. 2 (2007): 15–29.
5. Michele Betsill, "Mitigating Climate Change in US Cities: Opportunities and Obstacles," *Local Environment* 6, no. 4 (2001): 393–406.
6. Bai, "Integrating Global Environmental Concerns"; Rachel M. Krause, "An Assessment of the Greenhouse Gas Reducing Activities Being Implemented in US Cities," *Local Environment* 16, no. 2 (2011): 193–211.
7. David Satterthwaite, "Cities' Contribution to Global Warming: Notes on the Allocation of Greenhouse Gas Emissions," *Environment and Urbanization* 20, no. 2 (2008): 539–549.
8. Alistair Hunt and Paul Watkiss, "Climate Change Impacts and Adaptation in Cities: A Review of the Literature," *Climatic Change* 104, no. 1 (2011): 13–49.
9. ICLEI, *Local Government Implementation of Climate Protection: Report to the United Nations* (Toronto: International Council for Local Environmental Initiatives, 1997).
10. Betsill, "Mitigating Climate Change"; Harriet Bulkeley and Michele Betsill, *Cities and Climate Change: Urban Sustainability and Global Environmental Governance* (New York and London: Routledge, 2003).
11. Betsill, "Mitigating Climate Change"; D. E. Peterson, E. Matthews, and M. Weingarden, *Local Energy Plans in Practice: Case Studies of Austin and Denver* (Golden, CO: National Renewable Energy Laboratory, 2011), http://www.nrel.gov/tech_deployment/state_local_activities/pdfs/50498.pdf.

12. Stephen M. Wheeler, "State and Municipal Climate Change Plans: The First Generation," *Journal of the American Planning Association* 74, no. 4 (2008): 481–496.

13. Michele Betsill and Harriet Bulkeley, "Looking Back and Thinking Ahead: A Decade of Cities and Climate Change Research." *Local Environment* 12, no. 5 (2007): 448.

14. Keila Szpaller, "John Birch Society Tied to ICLEI Protest at Missoula City Council," *Missoulian,* January 1, 2012.

15. Keila Szpaller, "Missoula City Council Passes Plan to Be Carbon Neutral by 2025," *Missoulian,* January 29, 2013.

16. Leslie Kaufman and Kate Zernike, "Activists Fight Green Projects, Seeing U.N. Plot," *New York Times,* February 3, 2012.

17. ICLEI USA, *ICLEI's Five Milestones for Climate Mitigation,* http://www.icleiusa.org/climate_and_energy/climate_mitigation_guidance/iclei2019s-five-milestones-for-climate-mitigation.

18. ICLEI USA, *US Local Sustainability and Climate Action Plans,* http://www.icleiusa.org/action-center/planning/List%200f%20US%20Sustainability%20and%20Climate%20Plans.pdf.

19. Jungah Bae and Richard Feiock, "Forms of Government and Climate Change Policies in US Cities," *Urban Studies* 50, no. 4 (2013): 776–788.

20. Carlo Aall, Kyrre Groven, and Gard Lindseth, "The Scope of Action for Local Climate Policy: The Case of Norway," *Global Environmental Politics* 7, no. 2 (2007): 83–101.

21. Rachel M. Krause, "The Motivations Behind Municipal Climate Engagement: An Empirical Assessment of How Local Objectives Shape the Production of a Public Good," *Cityscape: A Journal of Policy Development and Research* 15, no. 1 (2013): 125–141.

22. Bae and Feiock, "Forms of Government."

23. Rachel M. Krause, "The Impact of Municipal Governments' Renewable Electricity Use on Greenhouse Gas Emissions in the United States," *Energy Policy 47* (2012): 246–253.

24. Bae and Feiock, "Forms of Government"; Krause, "An Assessment": City of Kansas City, Missouri, *Climate Protection Plan, 2008,* http://marc.org/environment/airq/pdf/CP-Plan-7-16-08.pdf.

25. Dennis Murphey, "Kansas City, Missouri Climate Protection Plan July 2008," presentation to the EPA's State and Local Climate and Energy Program, http://www.epa.gov/statelocalclimate/documents/pdf/murphey_presentation_11-17-2011.pdf.

26. City of Kansas City, *Climate Protection Plan,* 8.

27. Murphey, "Kansas City, Missouri Climate Protection Plan."

28. The following articles empirically support these points: Sammy Zahran, Himanshu Grover, Samuel D. Brody, and Arnold Vedlitz, "Risk, Stress, and Capacity Explaining Metropolitan Commitment to Climate Protection," *Urban Affairs Review* 43, no. 4 (2008): 447–474; Rachel M. Krause, "Political Decision-Making and the Local Provision of Public Goods: The Case of Municipal Climate Protection in the US," *Urban Studies* 49, no. 11 (2012): 2399–2417.

29. Zahran et al., "Risk, Stress, and Capacity"; Krause, "Political Decision-Making."

30. Krause, "Political Decision-Making."

31. James C. Clingermayer and Richard C. Feiock, *Institutional Constraints and Policy Choice: An Exploration of Local Governance* (Albany: SUNY Press, 2001).
32. Elaine B. Sharp, Dorothy M. Daley, and Michael S. Lynch, "Understanding Local Adoption and Implementation of Climate Change Mitigation Policy," *Urban Affairs Review* 47, no. 3 (2011): 433–457; Krause, "Political Decision-Making."
33. Bae and Feiock, "Forms of Government."
34. Barry G. Rabe, *Statehouse and Greenhouse: The Emerging Politics of American Climate Change Policy* (Washington, DC: Brookings Institution Press, 2004); Henrik Selin and Stacy D. VanDeveer, "Political Science and Prediction: What's Next for US Climate Change Policy?" *Review of Policy Research* 24, no. 1 (2007): 1–27.
35. Chicago Climate Action Plan, http://www.chicagoclimateaction.org.
36. Ibid., 3.
37. Julia Parzen, *Lessons Learned: Creating the Chicago Climate Action Plan* (Chicago: City of Chicago Department of Environment, 2009), http://www.chicagoclimateaction.org/filebin/pdf/LessonsLearned.pdf.
38. US Environmental Protection Agency, *Inventory of US Greenhouse Gas Emissions and Sinks: 1990–2011*, EPA 430-R-13–001 (Washington, DC: Environmental Protection Agency, 2013).
39. Aall et al., "Scope of Action."
40. Jonathan B. Wiener, "Think Globally, Act Globally: The Limits of Local Climate Policies," *University of Pennsylvania Law Review* 155, no. 6 (2007): 1961–1979.
41. Nicholas Lutsey and Daniel Sperling, "America's Bottom-Up Climate Change Mitigation Policy," *Energy Policy* 36, no. 2 (2008): 673–685.
42. This estimate is based on rough calculations that are available from the author.
43. US Environmental Protection Agency, Greenhouse Gas Equivalencies Calculator, http://www.epa.gov/cleanenergy/energy-resources/calculator.html#results.
44. Selin and VanDeveer, "Political Science and Prediction."
45. Kathleen E. Halvorsen, "Assessing the Effects of Public Participation," *Public Administration Review* 63, no. 5 (2003): 535–543.
46. City of New York, *PlaNYC: An Inventory of New York City Greenhouse Gas Emissions*, http://nytelecom.vo.llnwd.net/o15/agencies/planyc2030/pdf/greenhousegas_2012.pdf.
47. Ibid.
48. Kent E. Portney and Jeffrey M. Berry, "Participation and the Pursuit of Sustainability in US Cities," *Urban Affairs Review* 46, no. 1 (2010): 119–139.
49. Krause, "Motivations Behind."

PART II

Civic Society
and Climate Change

CHAPTER 5

Explaining Public Conflict and Consensus on the Climate

Dennis Chong

As MEDIA ATTENTION TO CLIMATE change increased dramatically in the past 25 years, so did public concern. But the problem has remained a relatively low priority for the American public despite grave warnings from scientists about the consequences of inaction. Furthermore, a scientific consensus on the risks of climate change has not prevented sharp divisions of public opinion reflecting partisan and ideological polarization on the issue among political leaders.

In this chapter, I discuss the changing state of public attitudes toward climate change since the issue entered public consciousness in the 1980s. During this period, the public has been exposed to ongoing debate on a complex scientific topic, with opposing sides providing contrasting interpretations of evidence and competing policy recommendations. The divergent perspectives represented in this debate have fractured the public. Whether these social and political differences can be bridged to enable action on policy solutions is one of the central political issues of our time.

In addressing this question, I evaluate attitudes, beliefs, and policy preferences, and I provide a model for explaining the dynamics of conflict and consensus. I also discuss the impact of scientific information, the economy, and the effects of framing on opinions. We will see that on the broad themes of climate change—its causes, the views of experts, the need for policy action—a plurality or majority of the public generally gets it right, in the sense of being on the same side as the scientific authorities. There are also favorable trends suggesting that knowledge of the effects of human action on the atmosphere and climate, although not sophisticated, have progressed significantly compared to past generations and that skepticism toward climate change will be harder to sustain among new generations that have a stronger environmental consciousness.

Although I focus on the state of public opinion, I will also emphasize that the obstacles to significant government action transcend shortcomings of public attention. An active conservative countermovement has attempted to reduce public concern over climate change, and the perpetuation of a business–environment dichotomy has hindered policy solutions. Achieving a more general public consensus on the issue requires both greater elite leadership on the status and implications of climate change and creative policies that reconcile competing values and economic interests underlying partisan and ideological polarization.

Consensus and Conflict

Over the course of the last century, there have been many significant shifts in public assumptions about the relationship between human actions and the climate. Scientists were already exploring in the late 1800s the idea that industrial production of CO_2 could gradually warm the Earth's atmosphere and affect the climate. By the middle of the 20th century, people had become increasingly conscious of the capacity of humans to alter and disrupt nature. Humans were no longer regarded to be merely in thrall of nature; they could significantly change the environment around them, especially through pollution of the atmosphere and water.[1] In the 1950s, scientists began to warn of the effects of CO_2 on the earth's climate and developed a reliable method to measure the amount of CO_2 in the atmosphere, which produced disturbing evidence of steadily increasing levels.

Mounting concern about CO_2 emissions and global warming in the 1980s factored into energy policy discussions and the 1990 Clean Air Act, as scientists testified to Congress about the dire consequences of inaction. Early climate models were used to forecast how global temperatures would rise if CO_2 concentrations continued to increase, leading to potentially catastrophic outcomes. In 1988, global warming was featured on the front page of the *New York Times,* with a report on testimony by James Hansen (head of NASA's Goddard Institute for Space Studies at the time) to a congressional hearing that climate change was almost certainly caused by greenhouse gas emissions and that its effects would be measurable within the next decade. The unusually hot summer of 1988 and other weather-related events produced a spike in news stories about global warming, further increasing public awareness of the concept. Also in 1988, the Intergovernmental Panel on Climate Change (IPCC) was established by the United Nations to gather and clarify scientific information about the climate, and it quickly became the authoritative voice on climate change among scientists.

Then as now, the priority given to global warming in the media fluctuated with events, especially the state of the economy and party politics. The frequency of media stories in newspapers and magazines and on television all peaked between 1989 and 1990 before dropping off in the early 1990s.[2]

In the 1990s, a shift from determining whether global warming was happening to deciding how to deal with the problem turned climate change into a political and economic issue. The George H. W. Bush administration moved cautiously and expressed reservations about the scientific evidence and the economic consequences of international agreements to reduce greenhouse gas emissions. Business interests represented by organizations such as the Global Climate Coalition fought back against the growing consensus on global warming by challenging the science and emphasizing the costs of solutions, claiming policies would increase energy prices, slow economic growth, and force the United States to bear greater costs than the developing world.[3]

Consciousness, Concern, and Causes

The political and economic divisions that emerged in the 1990s became manifest in public opinion and have persisted in current public attitudes toward climate change. Data gathered over the last 30 years allow us to track multiple dimensions of attitudes toward climate change: awareness of the issue, comprehension and knowledge, perceptions of scientists and scientific evidence, concern about the consequences of global warming, priorities between environmental protection and economic growth, and attitudes toward policy measures and specific international agreements.[4]

The development of public consciousness of global warming can be traced directly to the amount of media coverage of the issue in the mid to late 1980s. Public awareness grew as the summer of 1988 brought record temperatures to many parts of the United States and generated discussion of changes in the climate. By the mid-1990s, awareness of the concept of the *greenhouse effect* had diffused to about 80 percent of the population; today, recognition of this idea surpasses 90 percent. Gallup polls summarized in Figure 5.1 show that about two-thirds of the public prior to the 2008 economic crisis worried a great deal or a fair amount about global warming. Somewhat higher proportions registered in Pew national surveys—ranging from two-thirds to three-quarters of the public—consider global warming to be a very or somewhat serious problem.

Attribution of the causes of global warming has been a more contentious issue. As shown in Figure 5.2, Gallup polls over the past decade have found that about 50–60 percent of the public believes that global warming is due more to human activities (anthropogenic causes) than to natural temperature fluctuations. This segment decreased about 10 percent during the post-2007 recession (the absolute levels vary by polling organization, but the trend appears consistent across organizations), but has since recovered partly to earlier levels. Since 2009, the public has become more likely to trace the primary cause of global warming to human actions, although there are significant ideological differences in belief, with conservatives

Figure 5.1   Personal Worry About Global Warming, 1989–2013

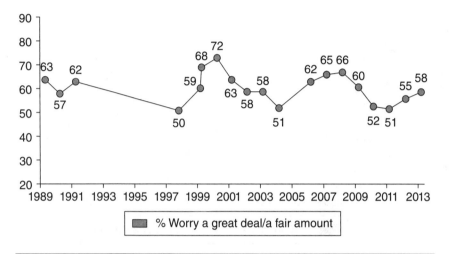

I'm going to read you a list of environmental problems. As I read each one, please tell me if you personally worry about this problem a great deal, a fair amount, only a little, or not at all. First, how much do you personally worry about [global warming]?

Source: Lydia Saad, "Americans' Concerns About Global Warming on the Rise," *Gallup Politics* (April 8, 2013), http://www.gallup.com/poll/161645/americans-concerns-global-warming-rise.aspx.

far more likely to point to natural fluctuations in temperatures as the source of warming (see below).

Agreement that global warming is occurring is also different from belief in the proximity of damages. Although scientists have been emphasizing the imminent effects of climate change, the public is inclined to think the problem will affect other areas and nations more than their own and that the effects will strike future generations more than themselves. Figure 5.3 shows that about two-thirds of the public believes that global warming will not pose a serious threat in their lifetimes. Since Gallup began asking this question in 1998, the proportion believing in an imminent threat has increased, but it has never exceeded 40 percent.

Global warming is not the highest priority of the public, even within the general class of environmental issues. For a majority of the public, environmental concerns in turn fall below concerns about the economy, health care, and US involvement in foreign wars. Among environmental issues, the priority placed on global warming trails water pollution—specifically concern about the purity of

FIGURE 5.2   Belief in Human vs. Natural Causes of Global Warming, 2003–2013

And from what you have heard or read, do you believe increases in the Earth's temperature over the last century are due more to — [the effects of pollution from human activities (or) natural changes in the environment that are not due to human activities]?

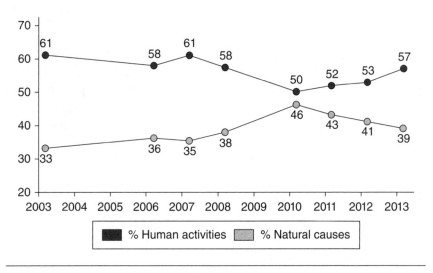

Source: Saad (2013).

drinking water, streams, rivers, and lakes. The public may believe that global warming will cause weather-related natural disasters with greater frequency, but the location of those disasters is generally thought to be remote in time and geography. (This may soon change in the aftermath of recent severe hurricanes Katrina and Sandy, which caused enormous destruction in highly populated metropolitan areas.)

Knowledge

Awareness of global warming does not imply or require scientific understanding. Objective knowledge is challenging to acquire given the scientific dimensions of the issue. Many people claim fair or moderate knowledge of the issue, while less than 30 percent of the public (in Gallup polls) claims to understand climate change "very well." This still represents about a 10 percent increase from eight to ten years ago, which suggests growing comprehension of the issue.

Whether we consider the public to be well or poorly informed naturally depends on our point of reference for making such evaluations. A 2012 knowledge

Figure 5.3   Proximity of the Threat of Global Warming, 1997–2013

Do you think that global warming will pose a serious threat to you or your way of life in your lifetime?

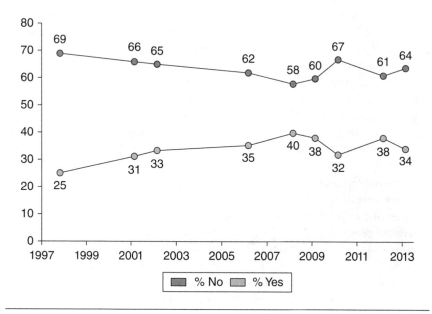

Source: Saad (2013).

survey, originating out of the Yale Project on Climate Change Communication, asked a broad set of questions testing knowledge levels in the general public. Although the researchers gave low letter grades to the public based on absolute scores (percentage correct), today's public is surely far more knowledgeable about climate, energy, and environmental issues than was the population in the decades prior to the emergence of global warming as an issue.

The Yale survey shows the American public to be reasonably informed about basic issues related to climate change, such as the connection between greenhouse gases and global warming, the effect of burning fossil fuels on carbon dioxide levels, the identification of coal and oil as major fossil fuels, the regional variation in effects of global warming on precipitation and temperatures, and the increased rate of melting of glaciers. Even though only about 50 percent say that global warming is mostly caused by human activities, large majorities agree that cars and trucks and burning fossil fuels for heat and electricity contribute a lot or some to global warming. A small plurality of 26 percent even correctly answered that global sea levels rose six to nine inches in the 20th century. When respondents were asked to estimate the average temperature of the Earth's surface 150 years ago (relative to its current average temperature of 58 degrees Fahrenheit), the median response of the public was 54 degrees, not significantly different from the IPCC estimate of between 56 and 57 degrees!

The public also accurately detects the convergence of scientific opinion on the problem of global warming. From 1994 through 2006, the proportion of the public that believed there was a scientific consensus that global warming was occurring more than doubled, from the mid–20 percent range to over 60 percent. Almost no one thinks most scientists dismiss or discount global warming. Instead, the remaining 40 percent of the public believes that scientists are divided on this question. Ideology is a major divider of perceptions, but conservatives and liberals are not diametrically opposed about the degree of scientific consensus.[5] Liberals are far more likely to say that scientists agree that the Earth is warming and that humans are responsible. Conservatives do not claim that scientists generally believe otherwise but instead maintain that scientists are "divided," which suggests they know the scientific majority disagrees with their own position (i.e., there is a limit to their motivated reasoning). A good test would be to ask people, "What percentage of scientists believe that global warming has been affected significantly by human behavior?" This would avoid the ambiguous meaning of *divided*, which could mean a 50–50 split or a 95–5 split.

A positive sign that augers well in the long run for the consensus view of global warming is that the carriers of that message—namely scientists—are members of one of the most trusted professions in society. According to a 2008 Yale/George Mason poll, climate scientists are trusted (strongly or somewhat) by 74 percent of respondents, a level exceeded only by other science organizations. Only a quarter of the public says they have little or no trust in scientists regarding the environment. By comparison, general distrust of politicians, religious leaders, and the media is far higher.

Generational Change

Another dynamic that foreshadows public opinion trends is the demographics of believers and skeptics. Age cohort differences are evident in the data on climate change opinion—young people are significantly more likely to believe in global warming and support environmental protection. A 2012 Gallup report identified large age cohort differences in priority assigned to economic growth and environmental protection (see Table 5.1). Among respondents 18–29, priority is given to the environment by 53 to 35 percent. All other age cohorts give priority to the economy, but the margin favoring the economy increases with age, with a 27 to 62 percent distribution among those 65 and older.

TABLE 5.1   Priority for Economic Growth or Environmental Protection by Selected Demographic Groups

Higher Priority for Economic Growth or Environmental Protection, by Subgroup, 2012		
	%Environment	%Economy
Republicans	27	66
Independents	44	43
Democrats	50	42
18 to 29 years	53	35
30 to 49 years	43	50
50 to 64 years	37	51
65+ years	27	62
Conservatives	29	64
Moderates	48	41
Liberals	56	35

Source: Jacobe (2012).

The generational changes in attitudes toward climate change mirror the dynamics of opinion change on controversial social issues. Pew studies show a broad-based generation gap in attitudes toward the role of government, trust in government, and opinions on specific issues such as gay rights, immigration, environmental regulation, and global warming. The tide is turning against the opposition on all of these issues. A 2011 Pew survey shows a generational divide especially between millennial (18–30-year-olds) and silent generation (66–83-year-olds) Americans in both their belief in global warming and their belief about the human causes of warming. But this generational conflict is exclusive to the Republican Party (see Figure 5.4). Consequently, there are partisan differences within every age cohort, but the ideological conflicts are smaller in the youngest cohort. Among millennial Republicans, almost 50 percent believe there is "solid evidence" the Earth is warming, whereas 75 percent of millennial Democrats are convinced that global warming is real. By comparison, the partisan gap is 41 percent in the silent generation. Such cohort differences are precursors of

Figure 5.4   The Partisan Divide on Climate Change by Age Cohort

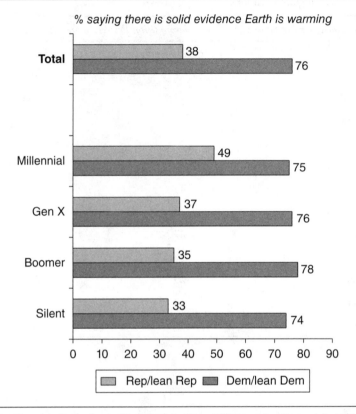

% saying there is solid evidence Earth is warming

Source: Pew Research Center for the People and the Press, *The Generation Gap and the 2012 Election* (Washington, DC: Pew Research Center, November 3, 2011), http://www.people-press .org/2011/11/03/section-8-domestic-and-foreign-policy-views/.

social change and portend the gradual development of a popular consensus on the issue.

Policy Preferences

The public's division over the cause of global warming does not preclude generally higher levels of agreement and support for a variety of policy measures aimed at addressing the problem. We can discern from responses to policy questions that people have been warned sufficiently about global warming to express concern, and many have gotten the message that society needs to invest more heavily in renewable energy sources and depend less on fossil fuels. The public also is quite willing in the abstract to say they would pay for reduced dependence on fossil fuels, but this willingness declines when concrete costs are mentioned. The behavioral intentions expressed by the public are not meaningless—indeed they may reflect changing attitudes that predict eventual behavioral changes—but they are largely symbolic expressions of support for concepts—clean energy, sustainability, energy efficient cars—that have been discussed favorably in public forums. Nonetheless, the clear direction of public sentiment on many of these issues shows that environmental consciousness is socially desirable, and there is potentially a large reservoir of support for action on climate change policies.

For example, a 2012 national survey shows that sizable majorities feel that Congress, corporations, and citizens should all be doing more to address global warming. Public opinion on more concrete policy proposals indicates majorities of varying sizes (from slightly over half to three-quarters) that support developing clean energy, research on renewable energy sources (73 percent), tax incentives to purchase energy efficient vehicles and solar panels (73 percent), and substituting renewable energy for fossil fuels.[6] (See also Chapter 7 on public support for climate change policies through initiatives and referendums.)

These results are consistent with repeated Gallup and Pew surveys on policy preferences showing large majority support for higher fuel efficiency standards, mandatory controls on CO_2 emissions by businesses, stricter standards on auto emissions, and the like. There are, however, also apparent contradictions in public attitudes on the issue, as majorities simultaneously call for increased exploration for and development of oil, coal, and gas out of a desire for independence from foreign sources of energy.[7]

In general, public enthusiasm for policy action tapers when costs are presented. Measures that involve higher taxes on gasoline or electricity, increased airfares, or increased taxes to subsidize clean energy receive tepid support. For example, in the 2012 national survey cited above, two-thirds (68 percent) of the public claim to support moderate- to large-scale efforts to reduce global warming, even if such actions require corresponding moderate to large costs. But only 10 percent strongly support a carbon tax on companies that produce or import fossil fuels if that would result in a $180 annual cost to the average household.

Many policy questions are undoubtedly only partially comprehended by respondents and constitute "doorstep opinions," referring to when household surveys more commonly were conducted in person, and people gave their opinions on issues they were possibly hearing about for the first time. For a number of years, surveys have included questions about cap-and-trade and carbon tax policies, with the results showing stronger support for cap-and-trade. However, when questioned in 2009, only 23 percent of the public knew that cap-and-trade was related to energy and environmental policy. An equal percentage thought cap-and-trade pertained to banking or health care, and almost half the sample admitted outright they did not know.

Attitudes and Actions

The public's motivation to take costly action will remain low until the problem is seen as impinging on people's everyday lives. In addition to being placed low in priority relative to other social and environmental problems, global warming has not been regarded as a looming personal threat compared to the perceived risk of heart disease and cancer. Public expressions of the threat posed by global warming refer more distantly or abstractly to societal problems that are removed from one's personal life.[8]

Abstract support for action, however, is strong in the United States, as it is across countries of the world.[9] But, as expected, studies in the United States and other European nations show that people's willingness to change their behavior or support a measure is inversely related to its cost or inconvenience.[10] For example, relatively high percentages of individuals claim they would drive or fly less frequently to reduce their contribution to carbon emissions, even if they would not pay higher gas prices or airfares. People are generally willing to make easy or inexpensive changes in their lifestyles to address climate change, but they discount the effect of changing their personal behavior and place most of the responsibility on government to take action.

The validity of public intentions to pay for policies that address climate change is difficult to ascertain. Responses may reflect the strength of people's attitudes on a subject rather than their economic evaluations. Indeed, studies have repeatedly found that willingness to pay (WTP) correlates with proenvironment attitudes and engagement, socioeconomic status, perceived efficacy of a policy, certainty of belief in climate change, expected future temperature and precipitation levels, and perceptions of other people's behavior.[11]

A Canadian review of research comparing data on people's expressed willingness in the abstract to pay for certain products (e.g., energy-saving home appliances) with consumers' actual behavior found wide discrepancies between the intentions conveyed in surveys and actual choice behavior in the marketplace.[12] A number of reasons can explain inflated WTP estimates. Individuals are less likely to behave as if they have a budget constraint when answering hypothetical questions. They may also want to please the interviewer or show they are generous or concerned about public goods. WTP scenarios also tend to exaggerate in the respondent's mind the connection between individual choices and social benefits. In real-world situations, where mutual cooperation is required, individuals may doubt that all parties will do their part to achieve a collective goal.

Complex WTP scenarios demand cognitive effort to understand what is being asked, and many respondents lack the skills or motivation required to provide reliable and valid responses.[13] Studies that present remote hypothetical scenarios involving, for example, reductions in emissions probably offer too little context for people to judge the tangible implications of these changes on the quality of their lives. Collecting valuation data may nevertheless help the policy-making process by identifying the most important features of environmental goods, the relevant trade-offs in considering a policy, the short- and long-term interests at stake, and the relevant audiences or stakeholders. Placing a monetary value on trade-offs may be superior to alternative approaches that use opinion measures of value trade-offs based on the direction and intensity of opinion.[14]

Berk and Fovell's paper is an exemplary study of how people respond sensibly to hypothetical but relevant changes in their local environments, in this case the Los Angeles metropolitan area. The design of their experiment ensured participants were more highly motivated (and capable) of answering questions about their willingness to pay for different outcomes. Respondents in warmer (Pasadena) and cooler (Marina Del Rey, Santa Monica, Malibu) areas of LA were presented with scenarios of climate change in their own areas using average temperatures and precipitation in those areas as baselines. Pasadena residents, for example, were presented with scenarios in which average summer temperatures climbed or fell, and average rainfall increased or decreased, and asked their willingness to pay for measures that would counteract or prevent these changes. In this manner, respondents were given realistic scenarios of climate shifts they could relate to, and the results identified which aspects of local climate change (more or less rain, higher or lower summer temperatures) were felt most acutely and which types of changes would prompt the greatest willingness to pay for preventive measures.

Explaining the Dynamics of Opinion Formation

Most discussions of public opinion on climate change assume a simple dynamic in which the public learns about the issue through media coverage and the messages communicated by opinion leaders through the media. The beliefs that people form depend on their evaluation of the stream of messages they receive. These are the key elements of the theory of "social learning" pioneered by McGuire (1968) and elaborated upon by Zaller's receive-accept-sample (RAS) model, in which the public takes its cues from trustworthy sources assumed to be informed about an issue and able to provide guidance in how to think about it.[15] According to this theory, people have different likelihoods of being exposed to any particular discussion of an issue. Furthermore, their acceptance of information is conditional upon the substance and cues of the message and the strength and direction of their political dispositions (i.e., their party identification and political ideology).

The key components in analyzing opinion formation are therefore the following:

- the strength, direction, and source of the message(s)
- the probability (P) of exposure (E) to that message (which is a function of political awareness or engagement with politics)

- the probability (P) of accepting (A) the message (which depends on the recipient's partisanship and recognition of the ideological or partisan cues contained in the message)

To see the dynamics of the model, consider an idealized model of the political environment:

Assume two political parties that send out messages (arguments) that are either similar or opposed to each other. Assume further that the members of the public have varying partisanship and degrees of attentiveness.

Scenario 1. Political elites are united in support of a policy. How does the public respond to the messages emanating from political elites through the media?

For any individual: P (adopting a belief) = P (exposure to an argument advocating the belief) x P (accepting the argument conditional upon exposure).

For Democrats, Republicans, and Independents, this means the probability of exposure will vary by their attentiveness (as indexed by interest, education, knowledge). The probability of acceptance will be high across the board, because the message has bipartisan support. If we assume that individuals look to their party leaders for cues (information shortcuts), then Democrats will look to Democratic leaders for guidance, and Republicans will look to Republican leaders. Independents will be more open to the views of both parties, but the message is the same no matter which way they turn. Everyone is persuaded to accept the message by his or her elite reference group.

When P(E) is multiplied by P(A given E), we find that support for the policy is a function of attentiveness, with more attentive individuals being the most supportive.

Scenario 2: Democratic leaders and Republican leaders take opposing positions on the policy.

Once again the P (exposure to elite messages) is correlated with attention to politics. But now we have mixed messages. For Democrats, the probability of acceptance of a Democratic message will be high, but probably higher among more knowledgeable Democrats, because they will more easily recognize that the message they have heard is indeed a Democratic message. Conversely the P (acceptance of a Republican message) among Democrats will decline with sophistication, because sophisticated Democrats know that they are receiving a partisan message from the opposing party. The same logic, adjusting for details, applies to the response of Republicans, leading to polarization between Democrats and Republicans with increasing exposure to elite communications. Independents are more likely to accept arguments originating from both sides, and will be most responsive to the relative loudness of the competing messages.

The applicability of the model to the climate change debate is apparent in Figure 5.5, based on a 2008 Pew national survey, which shows that college-educated Democrats and Republicans are further apart in their beliefs about the causes of global warming than are Democrats and Republicans who did not attend college. Education is a proxy for exposure to political debate. Therefore, Figure 5.5 shows that politically aware Republicans and Democrats are selectively accepting messages about global warming that are consistent with their predispositions. Partisans who are less educated are not as far apart, because they are receiving fewer polarizing messages and are less discriminating in their evaluation of those messages. Last, note that college-educated independents believe more strongly in the human causes of global warming, thus reflecting the stronger influence of the dominant scientific viewpoint in society.

Media Balance in the Flow of Information

The social learning model traces the dynamics of public opinion on any issue to the flow of information in society. In the mid- to late 1980s, global warming was not the established partisan and ideological issue that it became subsequently. Scientists were the predominant group speaking on the issue in the media, and their scientific reports became the basis of news stories that set the agenda. The confluence of a

FIGURE 5.5   Belief in the Cause of Global Warming by Party Identification and Education, 2008 Pew Survey

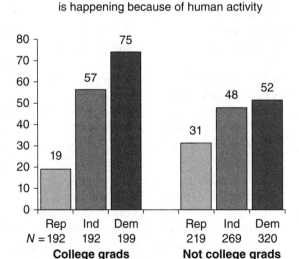

Source: Pew Research Center for the People and the Press, *A Deeper Partisan Divide Over Global Warming* (Washington, DC: Pew Research Center, May 8, 2008), http://www.people-press.org/2008/05/08/a-deeper-partisan-divide-over-global-warming/.

hot, dry summer across the nation in 1988 with media reports of that year greatly magnified attention to the issue.

The various media organizations influenced one another in their decisions to focus on climate change. Television coverage lagged behind print coverage, but television was a powerful catalyst in bringing the issue to public attention after it had been explored in greater detail in print. As the public became more concerned about climate change, the media stayed on the story.[16] However, the debate quickly shifted from the science of global warming to potential policy responses, as political and economic commentators moved to the forefront of the discussion in the media. Climate change became less of a scientific news story than a political story with political and economic coalitions beginning to take sides on the issue.

Despite the scientific consensus that global warming is caused largely by burning fossil fuels, the American public continues to be exposed to the minority views of contrarians or skeptics in politics, business, media, and policy organizations. The elite presses often balance their stories with opposing viewpoints that challenge the science of climate change, giving credence to the idea that climate change may be explained by naturalistic fluctuations rather than human action. Boykoff and Boykoff (2004) found that between 1988 and 2002 half of the news stories in several major newspapers gave comparable attention to the view that human actions were responsible for global warming and the contrary view that warming could be explained by natural variation of temperatures.[17] However, it is worth noting that a large proportion of stories coded in their study presented only the anthropogenic side, and hardly any stories represented the naturalistic side exclusively. Therefore, the scientific consensus predominated, but the contrarian position was sufficiently represented in stories to sustain and reinforce a solid minority oppositional view.

Ironically, media "objectivity"—in the sense of presenting opposing arguments without regard to their scientific status—gives audiences a distorted picture of reality, because it misrepresents the overwhelming majority view of experts on the subject. When there are competing streams of information, individuals tend to believe the sources they find more credible. In the terminology of psychology, audiences have to engage in more effortful "systematic" or "central route" processing of arguments to assess their validity, but they typically avoid such effort when the subject matter is difficult to comprehend or remote to their lives.[18] In the case of competing scientific claims, most individuals do not have the ability to evaluate the merit of opposing claims, but they often can evaluate the sources of those claims.

Criticism of the view that global warming is caused by humans has been more pronounced in the US media than in the UK media.[19] Dissenters from the scientific consensus have had greater influence in the United States than in the UK, perhaps because they have been affiliated with American think tanks and universities. The detractors have worked actively to promote a skeptical attitude toward

while weather changes and the quantity of scientific information about climate change (contained in science publications) were found to have little impact. The economic recession shifted attention away from climate change. Greater unemployment reduced perceptions of threat, while increases in GDP strengthened threat perceptions. Increases in war deaths in Iraq and Afghanistan also lowered perceptions of threat, suggesting competition among issues for public priorities.

Prior to 2008, elite cues on balance promoted greater concern about climate change, as Republican voting and statements against environmental measures decreased, and Democratic statements on climate change increased in quantity. Senator John McCain (R-AZ) worked with Democrats to produce legislation to combat climate change before he became the Republican presidential nominee. Al Gore's film, *An Inconvenient Truth,* also received most of its media attention prior to 2008. The subsequent decline in public concern can be traced to the increasingly partisan debate over climate change since 2008.

An additional detail in Scruggs and Benegal's analysis of the effects of the recession is that the public did not simply shift its priorities but also became less likely both to state that global warming was occurring and to link global warming to CO_2 emissions.[27] Therefore, the public seems to be rationalizing its concerns about the economy by questioning the reality of global warming. Furthermore, the relationship between unemployment rates and attitudes toward global warming are reproduced in public opinion in Europe, where there is less skepticism about global warming:

> A shift in the national unemployment rate from 5 to 9% in Europe (approximately the increase in unemployment in the United States during the time period) reduces the percentage of people reporting that global warming is a very serious problem by about 10 points.[28]

At the same time (despite declines in concern), surveys show there continues to be reasonably strong public support in the United States for taking actions to control carbon emissions, which indicates that general beliefs about climate change and support for policies can move independently. We may be paying excessive attention to whether the public recognizes the scientific consensus; the key issue is whether they support action, and they may support action irrespective of their beliefs about the causes of global warming.

Greater influences from weather changes were found by Donner and McDaniels using a measure of temperature variability based on the seasonal and monthly mean temperatures for 1990–2009.[29] Anomalies in temperature in the past 12 months were strongly correlated with beliefs in global warming. For every 1 degree Celsius increase in mean temperature, there was a 7.6 percent increase in the proportion of people who believed in global warming. An increase of

1 degree in the average temperature over the past 12 months led to a 10 percent increase in the proportion of the population that worried a great deal or a fair amount about global warming. Anomalies in seasonal temperatures are also highly correlated with the views expressed in the op-ed articles of major newspapers (*New York Times, Wall Street Journal, Washington Post, USA Today, Houston Chronicle*), which suggests that the media draws connections for the public between high temperatures and climate change. It is difficult to separate out the effects of coincidental events—in 2009, temperatures were colder than normal, but there was also an economic recession and the Climategate controversy over the communications of climate scientists. However, Donner and McDaniels speculate that less knowledgeable individuals with weaker attitudes toward climate change are more susceptible to being influenced by temperature fluctuations. They refer to these "swing voters" as being responsible for the short-term variation in the percentage concerned about climate change.[30]

Information Effects and Motivated Reasoning

Public opinion models start with the assumption that citizens will economize on the time they devote to political choices. The models allow little room for more careful consideration of arguments and their quality. Political arguments are not evaluated for their substance but according to their partisan or ideological sources. Consequently only a consensus among elites will swing opinion uniformly in support of a position. Divided elites produce divided publics.

The dynamics of public opinion on climate change shows that mere exposure to facts tends not to change policy positions when those positions are motivated by strong ideological values. Individuals with conservative leanings will maintain a contrarian position as long as it is championed by sources they find credible and trustworthy. Currently, both sides of the debate can cite elite sources and arguments for their positions, even though most scientists are in agreement.

The concept of motivated reasoning refers to the goals that give people incentives to engage in effortful thinking about a problem or issue. Kunda discussed two broad motivational goals behind reasoning: a desire to achieve a particular outcome (i.e., a directional goal), and a desire to be accurate or objective.[31] Directional motives lead people both to avoid and to discount evidence that runs counter to their prior attitudes and beliefs. In contrast, people may be motivated to be accurate when the issue is important to them (e.g., they may have an interest in making the right choice) or they have to explain or justify their positions to others. In such circumstances, they are more likely to try to be objective in their evaluation of evidence and to make their decision based on careful scrutiny of the credibility of sources and the strength of competing arguments.

Research on political communications shows that strong substantive arguments are able to prevail only when the partisan conflict is moderated. When there is a clear partisan divide, individuals tend to back their own party's position, even when that position is supported by weaker arguments. Druckman et al. (2013) found that individuals were influenced by the strength of arguments for and against oil drilling and the DREAM Act (about immigration reform) only when there were internal party divisions that diluted the partisan cues.[32] When the parties were represented as being polarized on the issue, individuals tended to side with their party and not be influenced by the strength of opposing arguments.

Indeed, Wood and Vedlitz's (2007) study of factors that influence whether people are concerned about global warming confirms that exposure to information about climate change in a nonpartisan context is more likely to be processed accurately.[33] Two survey experimental manipulations of information were tested. One manipulation varied the proportion of the public (40, 60, or 80 percent) that was said to be concerned about global warming. A second manipulation provided information stating that scientists have produced definite evidence of rising temperatures, melting icecaps, and rising sea levels but varied the number of degrees (either 2 or 5 degrees) the Earth's temperature would rise and the number of inches (6 inches or a foot) that sea levels would rise.

Both treatments had some impact on heightening concern for global warming. Varying the quantitative details in the second treatment had no further impact— perhaps not surprisingly, as people do not have an adequate reference point for evaluating more precise temperature and sea level quantities. Overall, these results suggest that social influence and factual information may affect individual priorities if they are not dominated by partisan cues. However, the information treatment was a one-sided message that did not account for the polarized atmosphere surrounding global warming. The problem facing efforts to raise concern for global warming is there is an active political opposition that seeks to muddy the scientific consensus and to undermine the credibility of scientists. If this experimental treatment were presented to a sample of strongly identified Republicans, the respondents would be more likely to counterargue against the claim and to discount its validity.

An example of directional bias in the evaluation of scientific evidence is a study by Sherman and Kunda involving scientific research on caffeine's effect on disease.[34] Some subjects read that caffeine facilitated progression of a serious disease; others read that caffeine inhibited progression. Caffeine drinkers were motivated to believe one conclusion over the other, and the reverse was true for low caffeine consumers. The results confirmed this directional bias. Threatened subjects were more likely to challenge the methodology of the research than non-threatened subjects.[35]

Elaborating on the caffeine study, Kunda writes,

Of importance is that all subjects were also quite responsive to the differential strength of different aspects of the method, which suggests that they were processing the evidence in depth. Threatened subjects did not deny that some aspects were strong, but they did not consider them to be as strong as did non-threatened subjects. Thus bias was constrained by plausibility. Taken together, these studies suggest that the evaluation of scientific evidence may be biased by whether people want to believe its conclusions. But people are not at liberty to believe anything they like; they are constrained by their prior beliefs about the acceptability of various procedures.[36]

These are hopeful results, showing that individuals viewing scientific evidence have prior beliefs about appropriate scientific methodologies that can restrict their ability to dismiss research results. People are constrained by their desire to provide explanations for their positions that would convince a "dispassionate observer."[37] Likewise, Pyszcynski and Greenberg speak of our desire to maintain an appearance of objectivity.[38]

By definition, the illusion of objectivity is shattered when no one believes you are seriously considering the available evidence on the issue. As long as Republican elites disagree with the scientific conclusions and can continue to enlist authorities to support their positions, rank-and-file members will fall in line. The more significant turning point will be when the skeptical position on climate change loses legitimacy and plausibility in public circles. It is clear that defenders of smoking cannot point to credible scientific evidence for their position. And it is becoming harder for those who doubt the existence of climate change to do the same. Max Boykoff's (2008) call for more responsible reporting of the status of scientific agreement and knowledge is aimed at reducing the legitimacy of politically motivated opposition:

Climate change is complex, which makes it a continually vexing problem for reporting, especially for a generalist audience. But in this high-stakes challenge, journalists and editors as well as scientists need to be intensely scrupulous. Media coverage should portray the contours of the varied aspects of climate change—from humans' role in it to whether it is "serious"—because better reporting has crucial implications for furthering understanding and potential public engagement. Granted, news will not provide the answer to climate change, but it does help to address, analyze and discuss the issues.[39]

Framing Public Policy More Effectively

The idea behind framing is that, on any issue, there are aspects or elements of the issue that are more likely to elicit favorable (or unfavorable) reactions from the public. If advocates or detractors emphasize those aspects of the issue, they become more accessible to the public, and the public is more likely to construe the issue using those selected frames of reference.

The most celebrated reframing of the global warming issue occurred when Republican pollster and campaign consultant Frank Luntz advised his clients to perpetuate the idea that climate science was inherently uncertain and to stop talking about global warming—which Luntz maintained had ominous implications— and instead to begin referring to "climate change." The term *climate change* had supposedly been found in focus group and polling research to represent a more neutral proposition than *global warming*. In fact, subsequent public opinion research (see below) has offered little support for this claim, but the example remains a vivid reminder of the putative influence of framing.

The use of metaphorical frames appears to have helped the public to comprehend the nature of several environmental problems: *acid rain,* the *ozone hole,* and the *hothouse* or *greenhouse effect* are metaphors that make issues accessible to nonexperts. These metaphors relate complex issues to processes or ideas that are familiar to the public. In past surveys (reviewed by Nisbet and Myers, 2007), it appears that when questions describe global warming as the greenhouse effect, respondents are more likely to claim they know something about this issue. The phrase *greenhouse effect* is accessible to nonscientists because people have felt the concentrated heat within a greenhouse. The ozone problem was effectively presented as a hole in a shield that needed to be repaired, which may have made it easier for people to visualize and become concerned about the problem. At the same time, confusion between the hole in the ozone layer and global warming perhaps also corresponds to people's intuitions. A hole in the ozone permits the rays of the sun to penetrate (and not be shielded by the ozone), which might be thought to increase the temperature of the Earth.

Certain claims are more readily acceptable because they are rooted in intuitively plausible stories.[40] When survey questions provide causal mechanisms for global warming—e.g., describing how carbon dioxide and other gases released into the atmosphere trap heat—respondents seem more likely to accept that this process will produce global warming. This illustrates how people need plausible stories to support their beliefs about the world. The greenhouse metaphor has become less salient over time in discussions of global warming, which may have inhibited self-reported knowledge of the problem, because it is a metaphor that makes the problem concrete and comprehensible for nonscientists.

Conversely, the counterclaim of skeptics that recent temperature increases are part of a natural cycle plays on the common belief that the environment is not human-made but part of the natural world and therefore generally outside the control of humans. Another common argument against climate change is that CO_2 is heavier than other air molecules (true) and will therefore fall to the surface rather than remain in the atmosphere (false). This false but intuitively plausible assertion therefore is a potentially effective frame for those who want to sow doubt about global warming. There is evidence the economy-versus-environment dichotomy can be overcome with the right plausible causal story. CBS and PIPA polls have shown that if the public is presented with the idea that efforts to curb greenhouse gas emissions will improve energy efficiency, respondents are more likely to say they believe such measures will help the economy by reducing energy costs.[41] Of course, these examples raise the intriguing question of where we get our intuitions about how the world works.

Experimental studies of the effect of framing on attitudes toward climate change have begun to identify approaches that may garner greater support for policies. Market-based frames used to describe policies appear to have greater appeal among individuals who give priority to hierarchical and individualist values over egalitarian and communitarian values.[42] Public health frames that emphasize connections between global warming and respiratory problems can make the effects of climate change more tangible and immediate to the public. "The emphasis also shifts the visualization of the issue away from remote arctic regions, peoples, and animals to more socially proximate neighbors and places such as suburbs and cities."[43]

Myers et al. argue that different ways of framing climate change make people more or less hopeful or angry, and in general elicit varying emotional responses from people.[44] They found a national security frame tended to anger those who are dismissive of climate change, while a public health frame tended to make those who were dismissive somewhat more hopeful. Although emotions were stirred by frames, there was no evidence that the environmental, national security, or public health frames could swing opponents around. None of this research however measured effects in a competitive context nor were any party cues presented alongside the frames.

Jon Krosnick compared how the public reacted to the alternative terms *climate change* versus *global warming* and increased "prices" versus increased "taxes" on gasoline in the United States and in Europe and found little difference, thus providing no support for the claims of consultants such as Frank Luntz.[45] Ironically, the term *climate change* elicited somewhat greater concern among Republicans and somewhat lesser concern among Democrats. Therefore, partisan polarization was reduced in reaction to climate change compared to global warming. Framing does appear to affect the priority placed on global warming. Yeager and colleagues (2011) have found that when the public is asked about the most important problem facing the world in the future (in contrast to the most important problem facing the nation today), global warming is more likely to be mentioned in that context.[46]

Research has identified conditions in which framing effects are likely to be mitigated. Of the potential antidotes to framing, the two most relevant in the context of the climate change issue are the effects of partisan communication sources and political competition. A frame—such as an argument that human actions are responsible for climate change—is most effective when it is received unopposed and without partisan cues. However the same message is far less effective when paired with an opposing frame—e.g., the argument that climate naturally varies over time. Whereas the recipients of one-sided frames are swayed by the considerations emphasized by that frame, the recipients of competing frames (i.e., frames representing alternative pro and con positions) are more likely to construe the issue using the frames that are congenial with their values.[47] Therefore, frames are dampened in politically polarized contexts. As noted earlier, when Democrats and Republicans are sending competing frames on an issue, partisan recipients tend to conform to the partisan position on the issue irrespective of the substance or strength of arguments made by each side.[48]

Reducing the Environment–Economy Conflict

A distinction must be drawn between debates over scientific claims and political conflict. Early disputes over the science of climate change gave way in the 1990s to debates over the economics and politics of national and international solutions. The political and economic debate in turn provided incentives for organized interests to challenge the science of global warming. Ultimately, disputes over interests are a powerful motive for sending conflicting messages to the public on the issue. Lessons can be drawn from Europe in promoting the economic benefits of policies aimed at addressing climate change. The contrast between United States and European public opinion is significant, because it highlights how the alignment of interests sharpens or diminishes the partisan divide.

Media coverage of the issue in Europe has promoted the idea of climate change to the point where skeptical views receive far less exposure than they do in the United States. Moreover, the European press has often been sensationalist in linking weather events such as floods, droughts, and heat waves to climate change without concern for scientific support.

Differences between the flow of information in the United States and Europe appear to be reflected in European public opinion. European Union citizens place a much higher priority on tackling climate change and see it as one of the most important global problems: two out of three respondents in 2011 said that climate change was among the most serious problems in the world, and 20 percent believed it was the world's single most serious problem. A majority ranked climate change as a more serious problem than current economic challenges following the 2008 financial crisis. Almost 80 percent said that policies to combat climate change would boost the economy and create jobs, an increase over 2009 when 63 percent agreed to the same statement.[49]

In Europe, the public faults business interests for not doing enough to address global warming. Ideological differences in Europe are more muted than they are in the United States. In a 2008 Eurobarometer survey, 66 percent of those who identified themselves as being on the "left" mentioned climate change as one of the most serious problems in the world, whereas 60 percent of those on the "right" did likewise.[50] When Europeans say that combating climate change will boost the economy, this is not a framing effect or a theoretical claim, as there is actual evidence for this relationship. In the United States, there is a persistent belief that the environment and economy must be traded off against each other.

Frames suggesting a positive connection between climate change policies and economic growth can help to shape public opinion, but more important to the success of such frames is whether business interests are persuaded they can profit from supporting new regulations. The claims contained in effective frames will not diminish elite opposition unless the major interests on the issue accept these claims. Until they do, they will continue to promote counterframes that sustain mass polarization.

Differences between Europe and the United States can be explained by different costs in adapting to lower emissions levels. European governments and businesses see benefits in reduced global emissions, because this would reduce energy demand and the price of energy. In Europe, the energy sector is weaker politically; generally more cooperative government–business relations on environmental policy have led to government regulations requiring changes in business practices that create incentives for businesses to adapt. US businesses have long felt they could successfully fight regulations, which caused businesses to fall behind in their ability to adapt to new environmental standards and, therefore, to be at a competitive disadvantage internationally. An important exception occurred in the campaign to ban chlorofluorocarbons (CFCs), where US business interests were more amenable than their counterparts in Europe, because they had a lower stake in the use of CFCs. US regulations had already forced American business interests to find alternatives, giving them an incentive to seek an international ban that would reduce the advantage enjoyed by European interests.[51]

North American businesses in the 1990s focused their energy on efforts to oppose regulation of greenhouse gas emissions. The Global Climate Coalition and the Climate Council argued against US support for the Kyoto Protocol. They also funded a major public relations campaign to discredit climate science and to argue that the economic costs of limiting greenhouse gases were excessive. The oil and automobile sectors were the most active business opponents along with other materials industries that relied heavily on fossil fuels.

The contrasting responses of businesses in the United States and Europe stem from their varying expectations about government regulation and their estimates of their ability to develop and profit from new technologies:

> Senior managers of European companies believed that climate change was a serious problem and that regulation of emissions was inevitable, but they were

optimistic about the prospects for new technologies. . . . American and Canadian companies, by contrast, tended to be more skeptical concerning the science, more pessimistic regarding the market potential of new technologies, and more confident of their political capacity to block regulation.[52]

Over time, the stance of US businesses has changed to recognize the issue of climate change is not disappearing and that they will need to factor it into their business models. Jones and Levy attribute some common movement of American and European countries toward accommodative strategies to be the result of global networks of senior managers, "which tend to induce similar expectations and norms concerning appropriate responses."[53]

American businesses continue to employ a mixed strategy by taking steps to adapt to regulations on carbon emissions but actively working to limit regulations that affect their profitability. In 2009 a well-funded effort by business succeeded in derailing the most significant climate policy bill—the Waxman–Markey American Clean Energy and Security Act. The cost of adapting to regulations on greenhouse gases should decline as technological innovations reduce the cost of compliance. Clean energy and low-emission technologies also will create new economic opportunities. But new regulations introduce uncertainties and risks in the marketplace that can be accepted or resisted. Businesses will be reluctant to change voluntarily if they believe they can delay the pace of change; more likely they will need to be pushed, as they have been in Europe by government action that has changed business incentives and strategies.

As businesses use public relations campaigns to persuade or inform the public of their actions to address climate change, critics contend that these actions are superficial and intended only to reduce public pressure. Nonetheless, we might expect that conciliatory messages originating from business sources such as British Petroleum and other oil companies will contribute to a growing sense of consensus on the need to address emissions and climate issues. At a minimum, these PR efforts represent a step in the process by which mainstream opinions begin to tip decisively to one side.

Conclusion: A Rational Public?

Public attitudes toward global warming conform to general theories of opinion that have been applied to other political issues. The public has a largely reactive role in these theories, as few political issues motivate ordinary citizens to pay more than passing attention to policy debates. The creators of public opinion are found among political elites—the politicians, media figures, journalists, intellectuals, interest groups, and activists that frame and shape debate and public opinion. Public opinion toward climate change therefore reflects patterns of elite communication over time.

A detached public can often make reasonable judgments by following elite cues, even if it lacks the motivation to appraise competing arguments and information on controversial issues. Overall, if we are evaluating the majority tendencies of the public, they are commonly on the correct side of factual issues. In the aggregate, majorities accurately perceive that the planet is warming, that human activities have played a significant role in contributing to the problem, and that most scientists believe that climate change is a real and dangerous phenomenon. There is fairly strong symbolic support for environmental ideas, concepts, and policies, although willingness to pay for climate change policies tapers off as their costs become more concrete and hit close to home. The American public on the whole mostly gets it right, even though virtually all attitudes and beliefs vary by political ideology and partisanship, with Republicans and conservatives being more skeptical about the science of climate change and more opposed to taking government action to address the problem.

Would the public hold the same attitudes and opinions about policy action if they learned the scientific facts and became more intensely involved with the issue? It is arguable a better-informed public might have a significantly different distribution of opinion, but to achieve such change, we require more responsible behavior from political elites and the media to accurately represent the state of knowledge about the issue, which is a different problem than an inattentive or apathetic public.

The public is responsive to the information flows of society.[54] When Democratic voices predominate on the issue, public concern about climate change increases. Conversely when partisan rancor grows, so do doubts about the reality and severity of the problem. Economic conditions influence both concern over climate change and beliefs about whether climate change is a real phenomenon. As the economy fell into recession in 2007, the public wanted government to deal with unemployment and other economic issues (the budget deficit, national debt, home foreclosures) that were felt to be more pressing concerns than climate change policy.

The worldwide economic downturn also made the public in the United States and Europe more skeptical that climate change was occurring and less inclined to attribute it to CO_2 levels, which is evidence of directionally motivated reasoning. Ignoring reality to suit one's desires is irrational, and individuals who do not properly evaluate information that has a bearing on their health may subject themselves to increased risks. The climate change debate belongs in this category as an issue on which motivated reasoning causes people to ignore, misinterpret or underweight real and present dangers in their environment. As Benedick warns,

> By the time the evidence on such issues as ozone layer depletion and climate change is beyond dispute, the damage could be irreversible and it may be too late to avoid serious harm to human life and draconian future costs to society.[55]

In the climate change debate, people are still prone to think climate change will have its greatest impact on future generations and regions other than their own. This long-term perspective is reinforced by contrarian efforts to undermine the validity of scientific arguments. Raising doubt about climate change reduces the relevance of the issue to people's lives and diminishes their motivation to seek and examine evidence on the subject.

However—and this should reassure supporters of government action—a large and growing proportion of the public acknowledges the reality of climate change and its damaging consequences, even if some have doubts about the relative contribution of human actions. Moreover, there are significant generational differences in attitudes, as younger individuals are more likely than older age cohorts to believe climate change is occurring and to endorse policies that address the problem. Generational differences exist because younger individuals are a barometer of the changing ideas and norms of society. The weaker predispositions of young people on the issue of climate change leave them open to recognizing the dominant flows of information in society. Even though competing viewpoints are represented in the media on climate issues, coverage favors the position taken by mainstream scientists. Few media stories present the skeptical view without qualification or opposition. The skeptical view is clearly the rearguard action in our society, and it appears to be losing out most decisively in the youngest generation of people, who also happen to see the world with the least bias.

Suggested Readings

Lorenzoni, I., and N. F. Pidgeon. "Public Views on Climate Change: European and USA Perspectives." *Climatic Change* 77, no. 1 (2006): 73–95.

McCright, Aaron M., and Riley E. Dunlap. "The Polarization of Climate Change and Polarization in the American Public's Views of Global Warming, 2001–2010." *Sociological Quarterly* 52 (2011): 155–194.

Nisbet, Matthew C. "Public Opinion and Political Participation." In *The Oxford Handbook of Climate Change and Society,* edited by J. Dryzek and R. Norgaard, 355–368. London: Oxford University Press, 2011.

References

American Enterprise Institute for Public Policy Research. *AEI Public Opinion Studies: Polls on the Environment, Energy, Global Warming, and Nuclear Power.* Washington, DC: American Enterprise Institute for Public Policy Research, 2013.

Bartels, L. M. "Beyond the Running Tally: Partisan Bias in Political Perceptions." *Political Behavior* 24, no. 2 (2002): 117–150.

Benedick, Richard E. *Ozone Diplomacy—New Directions in Safeguarding the Planet,* enlarged ed. Cambridge, MA and London: Harvard University Press, 1998.

Benedick, Richard E. "Science Inspiring Diplomacy: The Improbably Montreal Protocol." *Twenty Years of Ozone Decline*, edited by Christos Zerefos, Georgios Contopoulos, and Gregory Skalkeas, 13–19. Dordrecht: Springer, 2009.

Berk, Richard, and Robert G. Fovell. "Public Perceptions of Climate Change: A 'Willingness to Pay' Assessment." *Climatic Change* 41 (1998): 413–446.

Bord, Robert J., Ann Fisher, and Robert E. O'Connor. "Public Perceptions of Global Warming: US and International Perspectives." *Climate Change* 11 (1998): 75–84.

Borick, Christopher P. "American Public Opinion and Climate Change." In *Greenhouse Governance Addressing Climate Change in America*, edited by Barry G. Rabe, 24–57. Washington, DC: Brookings Institution Press, 2010.

Boykoff, Maxwell T. "The Real Swindle." *Nature Reports Climate Change*, February 21, 2008. doi:10.1038/climate.2008.14.

Boykoff, Maxwell T., and Jules M. Boykoff. "Balance as Bias: Global Warming and the US Prestige Press." *Global Environmental Change* 14 (2004): 125–136.

Boykoff, Maxwell T., and S. R. Rajan. "Signals and Noise: Mass-Media Coverage of Climate Change in the USA and the UK." *European Molecular Biology Organization Reports* 8, no. 3 (2007): 1–5.

Brulle, Robert J, Jason Carmichael, and J. Craig Jenkins. "Shifting Public Opinion on Climate Change: An Empirical Assessment of Factors Influencing Concern Over Climate Change in the U.S., 2002–2010." *Climatic Change*, February 6, 2012. doi:10.1007/s10584–012–0403-y.

Burstein, Paul. "The Impact of Public Opinion on Public Policy: A Review and an Agenda." *Political Research Quarterly* 56, no. 1 (2003): 29–40.

Chong, Dennis. *Rational Lives: Norms and Values in Politics and Society*. Chicago: University of Chicago Press, 2000.

Chong, Dennis, and James N. Druckman. "Dynamic Public Opinion: Communication Effects Over Time." *American Political Science Review* 104, no. 4 (2010): 663–680.

Chong, Dennis, and James N. Druckman. "Framing Public Opinion in Competitive Democracies." *American Political Science Review* 101, no. 4 (2007): 637–655.

Chong, Dennis, and James N. Druckman. "Identifying Frames in Political News." In *Sourcebook for Political Communication Research: Methods, Measures, and Analytical Techniques*, edited by Erik P. Bucy and R. Lance Holbert, 238–267. New York: Routledge, 2010.

Chong, Dennis, Herbert McClosky, and John Zaller. "Patterns of Support for Democratic and Capitalist Values." *British Journal of Political Science* 13 (1983): 401–440.

Cohen, G. L. "Party Over Policy: The Dominating Impact of Group Influence on Political Beliefs." *Journal of Personality and Social Psychology* 85, no. 5 (2003): 808–822.

Corner, Adam, Lorraine Whitmarsh, and Dimitrios Xenias. "Uncertainty, Skepticism and Attitudes Towards Climate Change: Biased Assimilation and Attitude Polarization." *Climatic Change* 114, no. 3–4 (2012): 463–478. doi:10.1007/s10584–012–0424–6.

Donner, Simon D., and Jeremy McDaniels. "The Influence of National Temperature Fluctuations on Opinions About Climate Change in the US Since 1990." *Climatic Change* 118, no. 3–4 (2013): 537–550.

Downs, Anthony. *An Economic Theory of Democracy*. New York: Harper and Row, 1957.

Druckman, James N., Erik Peterson, and Rune Slothuus. "How Elite Partisan Polarization Affects Public Opinion Formation." *American Political Science Review* 107, no. 1 (2013): 57–79.

Dunlap, Riley, and Aaron McCright. "A Widening Gap: Republican and Democratic Views on Climate Change." *Environment Magazine* (September–October, 2008). http://www.environmentmagazine.org/Archives/Back%20Issues/September-October%20 2008/dunlap-full.html.

Erikson, Robert S., Michael B. MacKuen, and James A. Stimson. *The Macro Polity.* Cambridge, UK: Cambridge University Press, 2002.

Eurobarometer. *Europeans' Attitudes Toward Climate Change.* European Commission, March-May, 2008.

Eurobarometer. *Climate Change.* European Commission, October, 2011.

Evans, Alex, and David Steven. *Climate Change: The State of the Debate.* London: The London Accord and Centre of International Cooperation, 2007.

Freudenburg, W. R. "Social Construction and Social Constrictions: Toward Analyzing the Social Construction of 'The Naturalized' as Well as 'The Natural.'" In *Environment and Global Modernity,* edited by G. Spaargaren, A. P. J. Mol, and F. H. Buttel, 103–119. London: Sage, 2000.

Hamilton, Lawrence C., Matthew J. Cutler, and Andrew Schaefer. "Public Knowledge and Concern About Polar-Region Warming." *Polar Geography* 35, no. 2 (2012):155–168.

Jacobe, Dennis. "Americans Still Prioritize Economic Growth Over Environment." *Gallup Economy,* March 29, 2012. http://www.gallup.com/poll/153515/americans-prioritize-economic-growth-environment.aspx.

Jones, Charles, and David Levy. "Business Strategies and Climate Change." In *Changing Climates in North American Politics,* edited by Henrik Selin and Stacy VanDeveer, 219–240. Cambridge MA: MIT Press, 2009.

Kahan, Dan M. "Cultural Cognition of Scientific Consensus." *Journal of Risk Research* (forthcoming).

Kahan, Dan M., Maggie Wittlin, Ellen Peters, Paul Slovic, Lisa Larrimore Ouellette, Donald Braman, and Gregory N. Mandel. *The Tragedy of the Risk-Perception Commons: Culture Conflict, Rationality Conflict, and Climate Change* (2011). Cultural Cognition Project Working Paper No. 89. New Haven, CT: Cultural Cognition Project, Yale University. http://dx.doi.org/10.2139/ssrn.1871503.

Kahneman, D. "Maps of Bounded Rationality: Psychology for Behavioral Economics." *American Economic Review* 93, no. 4 (2003): 1449–1475.

Kotchen, Matthew, Kevin Boyle, and Anthony Leiserowitz. "Willingness to Pay and Policy-Instrument Choice for Climate Change Policy in the United States." *Energy Policy* 55 (2013): 617–625.

Kunda, Z. "The Case for Motivated Political Reasoning." *Psychological Bulletin* 108, no. 3 (1990): 480–498.

Lau, Richard R., and David P. Redlawsk. *How Voters Decide: Information Processing During Election Campaigns.* Cambridge and New York: Cambridge University Press, 2006.

Lodge, M., and C. S. Taber. *The Rationalizing Voter.* New York: Cambridge University Press, 2013.

Lorenzoni, Irene, Anthony Leiserowitz, Miguel De Franca Doria, Wouter Poortinga, and Nick F. Pidgeon. "Cross-National Comparisons of Image Associations With 'Global Warming' and 'Climate Change' Among Laypeople in the United States of America and Great Britain." *Journal of Risk Research* 9, no. 3 (2006): 265–281.

Lorenzoni, Irene, and Nick F. Pidgeon. "Public Views on Climate Change: European and USA Perspectives." *Climatic Change* 77, no. 1 (2006): 73–95.

McCright, Aaron M., and Riley E. Dunlap. "Anti-Reflexivity: The American Conservative Movement's Success in Undermining Climate Science and Policy." *Theory, Culture & Society* 27, no. 2–3 (2010): 100–133.

McCright, Aaron M., and Riley E. Dunlap. "Defeating Kyoto: The Conservative Movement's Impact on U.S. Climate Change Policy." *Social Problems* 50, no. 3 (2003): 348–373.

McCright, Aaron M., and Riley E. Dunlap. "The Polarization of Climate Change and Polarization in the American Public's Views of Global Warming, 2001–2010." *Sociological Quarterly* 52 (2011): 155–194.

McGuire, William J. "Personality and Susceptibility to Social Influence." In *Handbook of Personality Theory and Research,* edited by E. G. Borgatta and W. W. Lambert, 1130–1187. Chicago: Rand McNally, 1968.

Myers, Teresa, Matthew Nisbet, Edward Maibach, and Anthony Leiserowitz. "A Public Health Frame Arouses Hopeful Emotions About Climate Change." *Climatic Change* 113, no. 3–4 (2012): 1105–1112.

Nemet, Gregory F., and Evan Johnson. "Willingness to Pay for Climate Policy: A Review of Estimates." La Follette School Working Paper No. 2010–011. Madison: University of Wisconsin, 2010. http://papers.ssrn.com/sol3/papers.cfm?abstract_id=1626931.

Nisbet, Matthew C. "Public Opinion and Political Participation." In *The Oxford Handbook of Climate Change and Society,* edited by J. Dryzek and R. Norgaard, 355–368. London: Oxford University Press, 2011.

Nisbet, Matthew C., and Teresa Myers. "The Poll-Trends: Twenty Years of Public Opinion About Global Warming." *Public Opinion Research* 71, no. 3 (2007): 444–470.

O'Connor, Robert E., Richard J. Bord, and Ann Fisher. "Risk Perceptions, General Environmental Beliefs, and Willingness to Address Climate Change." *Risk Analysis* 19, no. 3 (1999): 461–471.

Oliver, Fiona, and Danja van der Veldt. *Consumers' Willingness to Pay for Climate Change.* Final Report. Toronto, ON: Consumers Council of Canada, 2004.

Oreskes, Naomi, and E. Conway. *Merchants of Doubt.* New York: Bloomsbury, 2010.

Page, B. I., and R. Y. Shapiro. *The Rational Public: Fifty Years of Trends in Americans' Policy Preferences.* Chicago: University of Chicago Press, 1992.

Petty, R. E., and J. T. Cacioppo. "The Elaboration Likelihood Model of Persuasion." In *Advances in Experimental Social Psychology, edited by* L. Berkowitz, 123–205. New York: Academic, 1986.

Popkin, S. L. *The Reasoning Voter: Communication and Persuasion in Presidential Campaigns,* 2nd ed. Chicago: University of Chicago Press, 1994.

Pyszczynski, T., and J. Greenberg. "Toward and Integration of Cognitive and Motivational Perspectives on Social Inference: A Biased Hypothesis-Testing Model." In *Advances in*

Experimental Social Psychology, edited by L. Berkowitz, vol. 20, 297–340. New York: Academic, 1987.

Schuman, Howard, and Stanley Presser. *Questions & Answers in Attitude Surveys.* Thousand Oaks, CA: Sage, 1996.

Scruggs, Lyle, and Salil Benegal. "Declining Public Concern About Climate Change: Can We Blame the Great Recession?" *Global Environmental Change* 22, no. 2 (2012): 505–515.

Sherman, B. R., and Z. Kunda. " Motivated Evaluation of Scientific Evidence." Paper presented at the American Psychological Society convention, Arlington, Virginia, June 1989.

Sniderman, P. M., and S. M. Theriault. "The Structure of Political Argument and the Logic of Issue Framing." In *Studies in Public Opinion: Attitudes, Nonattitudes, Measurement Error, and Change,* edited by W. E. Saris and P. M. Sniderman, 133–165. Princeton, NJ: Princeton University Press.

Trumbo, Craig. "Constructing Climate Change: Claims and Frames in U.S. News Coverage of an Environmental Issue." *Public Understanding of Science* 5 (1996): 269–283.

Trumbo, Craig. "Longitudinal Modeling of Public Issues: An Application of the Agenda-Setting Process to the Issue of Global Warming." *Journalism and Mass Communication Monographs,* vol. 152. Columbia, SC: Association for Education in Journalism and Mass Communication, 1995.

Villar, Ana, and Jon A. Krosnick. "Global Warming vs. Climate Change Taxes vs. Prices: Does Word Choice Matter?" *Climatic Change* 105 (2011): 1–12.

Wood, B. Dan, and Arnold Vedlitz. "Issue Definition, Information Processing, and the Politics of Global Warming." *American Journal of Political Science* 51, no. 3 (2007): 552–568.

World Public Opinion. *30-Country Poll Finds Worldwide Consensus That Climate Change Is a Serious Problem.* Washington, DC: World Public Opinion, April 25, 2006. http:// www.worldpublicopinion.org/pipa/articles/btenvironmentra/187.php?lb=bte&.

World Public Opinion. *Most Would Pay Higher Energy Bills to Address Climate Change Says Global Poll.* Washington, DC: World Public Opinion, November 6, 2007. http://www .worldpublicopinion.org/pipa/articles/btenvironmentra/427.php?nid=&id=&pnt=427.

World Public Opinion. *Poll Finds Worldwide Agreement That Climate Change Is a Threat.* March 13, 2007. http://www.worldpublicopinion.org/pipa/articles/btenvironmentra/329 .php?nid=&id=&pnt=329.

World Public Opinion. *World Publics Willing to Bear Costs of Combating Climate Change.* Washington, DC: World Public Opinion, October 11, 2006. http://www.worldpublicopinion .org/pipa/articles/btenvironmentra/255.php?nid=&id=&pnt=255.

Yeager, David Scott, Samuel B. Larson, Jon A. Krosnick, and Trevor Tompson. "Measuring Americans' Issue Priorities: A New Version of the Most Important Problem Question Reveals More Concern About Global Warming and the Environment." *Public Opinion Quarterly* 75, no. 1 (2011): 125–138.

Zaller, John R. *The Nature and Origins of Mass Opinion.* Cambridge and New York: Cambridge University Press, 1992.

Notes

*My thanks to Yael Wolinsky-Nahmias for providing advice and comments on the writing of this chapter.

1. Alex Evans and David Steven, *Climate Change: The State of the Debate.* (London: The London Accord and Centre of International Cooperation, 2007).

2. Craig Trumbo, "Longitudinal Modeling of Public Issues: An Application of the Agenda-Setting Process to the Issue of Global Warming," *Journalism and Mass Communication Monographs,* no. 152 (Columbia, SC: Association for Education in Journalism and Mass Communication, 1995).

3. Evans and Steven, "Climate Change."

4. American Enterprise Institute for Public Policy Research (AEI), *AEI Public Opinion Studies: Polls on the Environment, Energy, Global Warming, and Nuclear Power* (Washington, DC: American Enterprise Institute for Public Policy Research, 2013). See also Matthew C. Nisbet and Teresa Myers, "The Poll-Trends: Twenty Years of Public Opinion about Global Warming," *Public Opinion Research* 71, no. 3 (2007): 444–470.

5. Dan M. Kahan, "Cultural Cognition of Scientific Consensus," *Journal of Risk Research* (forthcoming).

6. A. Leiserowitz, E. Maibach, C. Roser-Renouf, G. Feinberg, and P. Howe, *Public Support for Climate and Energy Policies in September, 2012.* New Haven, CT: Yale Project on Climate Change Communication, http://environment.yale.edu/climate/publications/Policy-Support-September-2012/.

7. AEI, *AEI Public Opinion Studies.*

8. Robert J. Bord, Ann Fisher, and Robert E. O'Connor, "Public Perceptions of Global Warming: US and International Perspectives," *Climate Change* 11 (1998): 75–84.

9. Evans and Steven, "Climate Change," 21; World Public Opinion, *30-Country Poll Finds Worldwide Consensus That Climate Change Is a Serious Problem* (Washington, DC: World Public Opinion, April 25, 2006).

10. Irene Lorenzoni and Nick F. Pidgeon, "Public Views on Climate Change: European and USA Perspectives," *Climatic Change* 77, no. 1 (2006): 73–95.

11. Gregory F. Nemet and Evan Johnson, *Willingness to Pay for Climate Policy: A Review of Estimates,* La Follette School Working Paper No. 2010–011. Madison: University of Wisconsin, 2010, http://papers.ssrn.com/sol3/papers.cfm?abstract_id=1626931; Matthew Kotchen, Kevin Boyle, and Anthony Leiserowitz, "Willingness to Pay and Policy-Instrument Choice for Climate Change Policy in the United States," *Energy Policy* 55 (2013): 617–625.

12. Fiona Oliver and Danja van der Veldt, *Consumers' Willingness to Pay for Climate Change.* Final Report (Toronto: Consumers Council of Canada, 2004).

13. Richard Berk and Robert G. Fovell, "Public Perceptions of Climate Change: A 'Willingness to Pay' Assessment," *Climatic Change* 41 (1998): 413–446.

14. Ibid.

15. John R. Zaller, *The Nature and Origins of Mass Opinion* (Cambridge and New York: Cambridge University Press, 1992); see also Dennis Chong, Herbert McClosky, and John Zaller, "Patterns of Support for Democratic and Capitalist Values," *British Journal of Political Science* 13 (1983): 401–440; Dennis Chong, *Rational Lives: Norms and Values*

in Politics and Society (Chicago: University of Chicago Press, 2000); William J. McGuire, "Personality and Susceptibility to Social Influence," in *Handbook of Personality Theory and Research*, ed. E. G. Borgatta and W. W. Lambert (Chicago: Rand McNally, 1968), 1130–1187.

16. Craig Trumbo, "Longitudinal Modeling of Public Issues: An Application of the Agenda-Setting Process on the Issue of Global Warning," *Journalism and Mass Communication Monograph*, vol. 152 (Columbia, SC: Association for Education in Journalism and Mass Communication, 1995).

17. Maxwell T. Boycoff and Jules M. Boykoff. "Balance as Bias: Global Warming and the US Prestige Press," *Global Environmental Change* 14: 125–136.

18. R. E. Petty and J. T. Cacioppo, "The Elaboration Likelihood Model of Persuasion," in *Advances in Experimental Social Psychology*, ed. L. Berkowitz (New York: Academic, 1986), 123–205.

19. Maxwell T. Boykoff and S. R. Rajan, "Signals and Noise: Mass-Media Coverage of Climate Change in the USA and the UK," *European Molecular Biology Organization Reports* 8, no. 3 (2007): 1–5.

20. Dennis Chong and James N. Druckman, "Identifying Frames in Political News," in *Sourcebook for Political Communication Research: Methods, Measures, and Analytical Techniques*, Erik P. Bucy and R. Lance Holbert (New York: Routledge, 2010), 253.

21. W. R. Freudenburg, "Social Construction and Social Constrictions: Toward Analyzing the Social Construction of 'The Naturalized' as Well as 'The Natural,'" in *Environment and Global Modernity*, G. Spaargaren, A. P. J. Mol, and F. H. Buttel (London: Sage, 2000), 103–119.

22. Z. Kunda, "The Case for Motivated Political Reasoning," *Psychological Bulletin* 108, no. 3 (1990): 480–498.

23. Aaron M. McCright and Riley E. Dunlap, "Defeating Kyoto: The Conservative Movement's Impact on U.S. Climate Change Policy," *Social Problems* 50, no. 3 (2003): 348–373; Naomi Oreskes and E. Conway, *Merchants of Doubt* (New York: Bloomsbury, 2010).

24. Aaron M. McCright and Riley E. Dunlap, "The Polarization of Climate Change and Polarization in the American Public's Views of Global Warming, 2001–2010," *Sociological Quarterly* 52 (2011): 155–194.

25. Riley Dunlap and Aaron McCright, "A Widening Gap: Republican and Democratic Views on Climate Change, *Environment Magazine* (September–October, 2008), http://www.environmentmagazine.org/Archives/Back%20Issues/September-October%202008/dunlap-full.html.

26. Robert J. Brulle, Jason Carmichael, and J. Craig Jenkins, "Shifting Public Opinion on Climate Change: An Empirical Assessment of Factors Influencing Concern Over Climate Change in the U.S., 2002–2010, " *Climatic Change*, February 6, 2012, doi:10:1007/s10584–012–0403-y.

27. Lyle Scruggs and Salil Benegal, "Declining Public Concern About Climate Change: Can We Blame the Great Recession?" *Global Environmental Change* 22, no. 2 (2012): 505–515.

28. Ibid., 513.

29. Simon D. Donner and Jeremy McDaniels, "The Influence of National Temperature Fluctuations on Opinions About Climate Change in the U.S. Since 1990," *Climatic Change* 118, nos. 3–4 (2013): 537–550.

30. Ibid.

31. Kunda, "Case for Reasoning."

32. James N. Druckman, Erik Peterson, and Rune Slothuus, "How Elite Partisan Polarization Affects Public Opinion Formation," *American Political Science Review* 107, no. 1 (2013): 57–79.

33. B. Dan Wood and Arnold Vedlitz, "Issue Definition, Information Processing, and the Politics of Global Warming," *American Journal of Political Science* 51, no. 3 (2007): 552–568.

34. B. R. Sherman, and Z. Kunda, " Motivated Evaluation of Scientific Evidence," paper presented at the American Psychological Society convention, Arlington, Virginia, June 1989.

35. For a comparable experiment using scientific information about climate change, see Adam Corner, Lorraine Whitmarsh, and Dimitrios Xenias, "Uncertainty, Skepticism and Attitudes Towards Climate Change: Biased Assimilation and Attitude Polarization," *Climatic Change* 114, no. 3–4 (2012): 463–478, doi:10.1007/s10584–012–0424–6.

36. Kunda, "Case for Reasoning."

37. Ibid.

38. T. Pyszczynski and J. Greenberg, "Toward an Integration of Cognitive and Motivational Perspectives on Social Inference: A Biased Hypothesis-Testing Model," in *Advances in Experimental Social Psychology,* ed. L. Berkowitz, vol. 20 (New York: Academic, 1987), 297–340.

39. Maxwell T. Boykoff, "The Real Swindle," *Nature Reports Climate Change,* February 21, 2008, doi:10.1038/climate.2008.14.

40. D. Kahneman, "Maps of Bounded Rationality: Psychology for Behavioral Economics," *American Economic Review* 93, no. 4 (2003): 1449–1475.

41. These are my observations based on the data collected in Matthew C. Nisbet and Teresa Myers, "The Poll-Trends: Twenty Years of Public Opinion About Global Warming," *Public Opinion Research* 71, no. 3 (2007): 444–470.

42. Dan M. Kahan, "Cultural Cognition of Scientific Consensus," *Journal of Risk Research* (forthcoming).

43. Matthew C. Nisbet, "Public Opinion and Political Participation," in *The Oxford Handbook of Climate Change and Society,* J. Dryzek and R. Norgaard (London: Oxford University Press, 2011), 362.

44. Teresa Myers, Matthew Nisbet, Edward Maibach, and Anthony Leiserowitz, "A Public Health Frame Arouses Hopeful Emotions about Climate Change," *Climatic Change* 113, no. 3–4 (2012): 1105–1112.

45. Ana Villar and Jon A. Krosnick, "Global Warming vs. Climate Change Taxes vs. Prices: Does Word Choice Matter?" *Climatic Change* 105 (2011): 1–12.

46. David Scott Yeager, Samuel B. Larson, Jon A. Krosnick, and Trevor Tompson, "Measuring Americans' Issue Priorities: A New Version of the Most Important Problem Question Reveals More Concern About Global Warming and the Environment," *Public Opinion Quarterly* 75, no. 1 (2011): 125–138.

47. Dennis Chong and James N. Druckman, "Framing Public Opinion in Competitive Democracies," *American Political Science Review* 101, no. 4 (2007): 637–655; Chong and Druckman, "Identifying Frames in Political News"; P. M. Sniderman and S. M. Theriault, "The Structure of Political Argument and the Logic of Issue Framing," in *Studies in Public Opinion: Attitudes, Nonattitudes, Measurement Error, and Change*, ed. W. E. Saris and P. M. Sniderman (Princeton, NJ: Princeton University Press, 2004), 133–165.

48. James N. Druckman, Erik Peterson, and Rune Slothuus, "How Elite Partisan Polarization Affects Public Opinion Formation," *American Political Science Review* 107, no. 1 (2013): 57–79. See also G. L. Cohen, "Party Over Policy: The Dominating Impact of Group Influence on Political Beliefs," *Journal of Personality and Social Psychology* 85, no. 5 (2003): 808–822.

49. Eurobarometer, *Climate Change* (European Commission, October, 2011).

50. Eurobarometer, *Europeans' Attitudes Toward Climate Change* (European Commission, March-May, 2008).

51. Richard E. Benedick, *Ozone Diplomacy—New Directions in Safeguarding the Planet*, enlarged ed. (Cambridge, MA, and London: Harvard University Press).

52. Charles Jones and David Levy, "Business Strategies and Climate Change," in *Changing Climates in North American Politics*, ed. Henrik Selin and Stacy VanDeveer (Cambridge, MA: MIT Press), 223.

53. Ibid.

54. B. I. Page and R. Y. Shapiro, *The Rational Public: Fifty Years of Trends in Americans' Policy Preferences* (Chicago: University of Chicago Press, 1992); Robert S. Erikson, Michael B. MacKuen, and James A. Stimson, *The Macro Polity* (Cambridge, UK: Cambridge University Press, 2002).

55. Benedick, *Ozone Diplomacy*, 20.

CHAPTER 6

The US National Climate Change Movement

Robert J. Brulle

> Most other environmental problems pale beside the implications
> of a major climate shift.
>
> —Sheldon Kinsel, National Wildlife Federation, June 8, 1977[1]

CLIMATE CHANGE ENTERED into the political arena in 1977. In a congressional hearing on the environmental implications of the Carter administration's energy plans, the representative of the National Wildlife Federation noted that the development of the proposed synthetic fuels program would lead to massive releases of carbon dioxide and destabilize the planet's climate. Since then, climate change has been propelled to the forefront of environmental concern. Accompanying this concern has been the rise of a social movement that has advocated for strong action to mitigate CO_2 emissions. In just 30 years, the climate change movement has grown from a small contingent of organizations that were involved in atmospheric environmental concerns into a major component of the US environmental movement.

How did this movement develop? What are the different viewpoints represented in this movement? What are the most influential groups? In this chapter, I provide a detailed empirical description of the national climate change movement in the United States. In the first part of this chapter, I provide an overview of the unique role that social movements play in the process of social change. I then detail the historical development of the US national climate change movement from 1980 to the present. Then, based on a sample of climate change coalitions, I discuss the different discursive frames that define this movement and the comparative organizational makeup of the different coalitions. I conclude with an analysis of the different levels of cultural and political influence among the organizations that are focused at the national level of US climate change policy.

Civil Society and Institutional Change

As the modern social order developed, the institutions of society formed into three distinct spheres: the market, the state, and civil society. The market sector developed from traditional barter exchanges into production and exchange relationships coordinated through the market. Conjointly, the government or state sector developed as a means of ensuring the operation of and stabilizing the effects of the economic system. Thus economic interaction and the social relations of production became coordinated through the steering mechanisms of money and power carried out in the institutions of the market and the state.[2] Action within these sectors is constrained within parameters defined by their key imperatives. For the market, this is the necessity to maximize return on investment through continuous economic expansion. For the state, this entails providing security, ensuring economic growth, and maintaining its political legitimacy.[3] Accordingly, environmental actions that impinge on any of these imperatives will not be fostered within the dynamics of the market or the state. Rather than transforming economic and political institutions to meet ecological limitations, this dynamic forces the creation of environmental policies to fit into the maintenance of existing institutions.[4] Thus our society's capability for self-correction is systematically limited by the institutions of the capitalist world economy and the nation state. This greatly restricts the range of possible policy considerations, such as global governance or moving from an economy centered on status consumption to providing for human satisfaction.

The third sector, civil society, has been identified as a key site for the origination of large-scale social change.[5] Civil society is constituted by voluntary institutions that exist outside of the direct control of both the market and the state. Because these organizations are nominally based in the deliberations of their members, the institutional dynamics defined by market forces and the political power of the state are minimized in their operation. This independence forms the key to the capacity of civil society to serve as a site for the generation of democratic action.[6] Since they are based in communicative action, the institutions of civil society constitute a means to identify and propose actions to resolve social and environmental problems, unhindered by the limitations of institutions based in either the market or the state.[7] This puts civil society at the center of the renewal and transformation of social institutions.[8]

A key action that originates in civil society for the promotion of social change is the formation of social movement organizations. A social movement organization enables individuals to join together with other members of their community to participate meaningfully in their own governance. Such efforts allow individual citizens to translate their everyday concerns into collective issues and then press the government and economic institutions to address them.[9] This is evident in the area of climate change politics. A host of social movement organizations have advocated for strong government action to mitigate CO_2 emissions.[10]

The Historical Development of the US National Climate Change Movement

The US environmental movement is perhaps the single largest, longest-running, and most differentiated social movement in the United States.[11] It contains a number of different coalitions, spanning a broad range of issues from the preservation of biodiversity to advocacy for renewable energy. For the most part, the US national climate change movement grew out of existing environmental organizations that expanded their areas of concern to include climate change. To identify the national component of the climate change movement, this chapter looks at three different dimensions: First, has the organization testified in a congressional hearing on climate change; second, has the organization been recognized and participated in the meetings organized under the auspices of the United Nations Framework Convention on Climate Change (UNFCCC); and third, has the organization participated in climate change coalition focused on influencing policy at a national or regional level. If the organization participated in any of these three different dimensions, it was counted as a US national climate change organization.

Appearances at Congressional Hearings on Climate Change

Climate change advocacy certainly surfaced in the congressional hearing held on June 8, 1977. In the 1970s, climate change came up as an auxiliary issue in congressional hearings that were focused on atmospheric environmental issues, including ozone depletion and acid rain. Organizations such as the National Wildlife Federation, the Natural Resources Defense Council, and Resources for the Future were involved in atmospheric issues, and so the transition to climate change was a logical extension of their environmental concerns. Perhaps the first congressional hearing on the issue of climate change occurred on April 3, 1980, when the Senate Committee on Energy and Natural Resources convened a hearing on the "Effects of Carbon Dioxide Buildup in the Atmosphere." This hearing marked the beginning of focused climate change hearings in Congress. Since then, Congress has held over 500 hearings on this topic.

As political interest in this topic increased, additional groups began to testify before Congress. At first, the number of groups grew slowly. However, after the dramatic testimony by Dr. James Hansen in the summer of 1988, political interest in the issue expanded, and the number of groups testifying before Congress increased. Another rapid increase followed the release of Al Gore's movie, *An Inconvenient Truth,* in 2006. By 2010, 123 environmental movement organizations (EMOs) had testified before Congress in hearings on climate change.

Attendance at United Nations
Framework Convention on Climate Change Meetings

This growth at the federal level was paralleled by EMO participation in the meetings associated with the procedures defined by the UNFCCC. Starting with the initial discussions of the makeup of this treaty in 1991, there was a steady annual increase of EMO participation in the annual Conference of the Parties (COP) meetings. Participation in the UNFCCC meetings underwent a dramatic increase in 2009. This was due to the increased importance attached to COP 15, held in Copenhagen.

Taken together, the number of organizations involved, either in providing congressional testimony regarding climate change or participating in the UNFCCC process, increased from only 7 organizations in 1980 to 240 organizations in 2010. This growth in EMOs advocating for climate change action is shown in Figure 6.1.

Formation of Coalitions to Address Climate Change

A third dimension of the historic growth of EMO advocacy on the issue of climate change is the growth of coalitions that are specifically focused on that issue. The formation of coalitions is a regular feature of social movements, and the national climate change movement is not unique in this regard. What is unique is the wide variety of coalitions, which is due to the diversity of the US environmental movement and to how different discursive communities have addressed this issue. An examination of these coalitions can show the formalized and regular patterns of cooperation among different environmental organizations and thus provide us an empirical map of the extent and boundaries of the national climate change movement.[12] Based on a comprehensive web search and conversations with leaders of the current climate change movement, 21 unique coalitions involved in the issue of climate change at the national or regional level were identified. A list of these coalitions, in order of founding year, is provided in Table 6.1.

As this table shows, the first coalition, the US Climate Action Network, was founded in 1989. The Sustainable Energy Coalition followed next in 1992, along with Via Campesina. There was then a six-year break in the mid-1990s until two additional coalitions were formed in 1998. However, as this table illustrates, over half of the coalitions were formed in the 2006–2009 time period. This corresponds to the increased growth of organizations participating in the UNFCCC process and appearing before Congress.

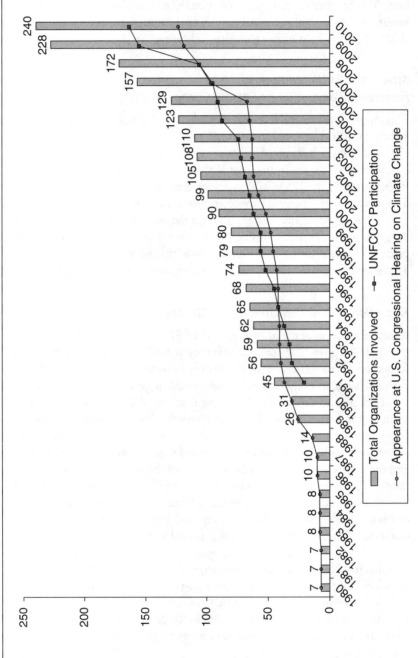

FIGURE 6.1   US Environmental Organization Participation in Climate Change Forums

TABLE 6.1   Environmental Movement Coalitions, by Year of Formation

Coalition name	Year founded
US Climate Action Network	1989
Sustainable Energy Coalition	1992
Via Campesina North American Region	1992
Chesapeake Climate Action Network (CCAN)	1998
Interfaith Power & Light Network	1998
Rising Tide North America	2000
Apollo Alliance	2001
Interwest Energy Alliance	2002
Climate Crisis Coalition	2004
Energy Action Coalition	2004
BlueGreen Alliance	2006
Catholic Coalition on Climate Change	2006
Interfaith Climate Change Network	2006
350.org/1Sky	2007
Rocky Mountain Climate Organization	2007
Season End	2007
US Climate Action Partnership	2007
Mobilization for Climate Justice	2008
TckTckTck/Global Campaign for Climate Action	2008
Clean Energy Works	2009
Climate SOS	2009

Overall, the pattern of growth of EMOs involved in climate change advocacy at the national or international level was one of gradual involvement up till about 2005. Then, from 2006 to 2009, there was a rapid increase in the number of organizations involved in either participating in formal government arenas or in coalition actions. Taken together, this pattern of involvement identifies 467 unique organizations that have either belonged to a climate change–focused coalition, participated in the UNFCCC process, or testified before Congress. While there may be some key EMOs that are not included in this sample, it provides a robust sample of the organizations that compose the core of the national US climate change movement.

Framing of Climate Change in the United States

Across these organizations, the cultural perspectives on the nature of and solutions to climate change vary dramatically. For example, the Environmental Defense Fund works with industry and trade associations to develop market mechanisms to promote the development of carbon-free energy technologies. Alternatively, Rising Tide North America aims at the transformation of capitalism and reductions in the overall consumption levels in the United States. Thus an understanding of the diversity of viewpoints across national climate change EMOs provides a critical dimension of our understanding of this movement.

For the past 25 years, social movement scholars have utilized Goffman's frame analytic perspective[13] to focus on the construction, maintenance, and alteration of social movement worldviews.[14] One of the central analytic foci has been the development of a movement's discursive frame.[15] A discursive frame provides a system of shared cultural beliefs that creates a network of action among culturally aligned organizations. Within any social movement, there are generally multiple frames, defining distinct movement sectors or "wings" that diverge in terms of their definition of problems, strategies, and methods of organization.[16] Thus a key to understanding the structure and dynamics of a social movement is the nature and distribution of frames within the movement. One approach to defining discursive frames within the climate change community was developed by Bäckstrand and Lövbrand.[17] In their analysis, there are three dominant discursive frames that define climate change discourse.

Climate Change Discourses at the Global Level

The first perspective is defined as *green governmentality*.[18] This discursive frame defines a process based on a scientific analysis of climate change and the development of global governmental initiatives to address this issue. The solution to climate change in the discourse of green governmentality is the implementation of a strong system of governance of the economy, natural resource use, and individual behavior informed by the natural sciences that is developed at an international level. In this approach, natural scientists play a key role in global environmental management[19] by mediating between science and politics through providing information to the public and decision makers in government and industry of the need for governmental action. This places scientists is the key role of defining the nature of this problem and proposing mechanisms for their resolution. In essence, this viewpoint legitimates the creation of an ecotocracy. This approach underlies many of the existing international treaty frameworks, in which science-based resource management plays a central role. It also informs actions aimed at the proximate causes of environmental degradation, such as creating parks or land trusts to preserve ecosystems or developing new technologies that can provide low carbon energy.

The second major discursive approach is known as *ecological modernization*.[20] This discursive frame focuses on the role of technological development, economic expansion, and the growth of environmental governance in creating and also mitigating environmental problems. In this perspective, economic development and shifts in technology lead to the initial generation of increased CO_2 emissions. However, further economic development can also mitigate these emissions by shifting to renewable energy technologies and energy efficiency.[21] This technological shift can thus result in a decrease in carbon emissions and a decoupling of economic growth and energy production with carbon emissions.[22] Thus in this perspective economic growth can result in an absolute decline in levels of environmental pollution.[23] At the core of ecological modernization theory is that existing social, economic, and governmental institutions can effectively deal with environmental issues, and there is no need for radical structural changes in industrial society.[24] This leads to a neoliberal market approach to the resolution of climate change.

This approach take two forms.[25] The first form is known as *weak ecological modernization*. This version focuses on technological development and energy efficiency, and it includes the use of market-based user fees for pollution, tax incentives, increases in energy efficiency, or the shifting of production toward green products.[26] The second version, *strong ecological modernization*, focuses on embedding environmental and ecological concerns in society through the reconfiguration of existing political and economic institutions. This includes adjustment of economic systems to include the value of natural capital in production decisions, modification of the existing political system toward more democratic participation, and inclusion of developing countries' social justice and equity concerns into global environmental governance.[27] Of these two approaches, weak ecological modernization is generally considered to be dominant.

Over the past decade, the discursive frames of global governmentality and ecological modernization have tended toward a merger. Since the viability of ecological modernization mechanisms, such as a global price on carbon, is dependent on the institution of a global accord, the two approaches to address climate change are seen as mutually constitutive. Together, these two perspectives form the dominant climate change discursive frame.

Distinct from these two perspectives is the discourse of *civic environmentalism*.[28] Civic environmentalism offers a counternarrative to the dominant climate change discourse. There are two related approaches within this discursive frame. The first perspective is defined by Bäckstrand and Lövbrand[29] as *radical resistance*. From this discursive perspective, both ecological modernization and global governance are seen as favoring the interests of the existing power elites and the dominant industrialized countries, and thus these discourses result in the marginalization of poor people and less-developed countries. Radical resistance challenges the neoliberal approach embedded in ecological modernization and calls for the radical democratization of global governance and economic processes. It aims at "a fundamental transformation of consumption patterns and existing institutions to realize a more eco-centric and equitable world order."[30] It focuses on the notion of global climate justice, emphasizing the responsibilities of the developed countries to dramatically reduce their carbon emissions, and the equitable sharing of technology and capital to enable the poorest nations to address global climate change. It also presses for the democratic reform of large multilateral institutions, such as the International Monetary Fund and the World Bank.

The second perspective within civic environmentalism is *reformist,* or *participatory multilateralism.*[31] This version of civic environmentalism focuses on opening up climate change treaty negotiations to wider participation by representatives of civil society. The argument is that becoming more inclusive will generate greater legitimacy for the negotiated agreements, and increase the implementation of actions that maximize the benefits from neoliberal approaches.

Climate Change Discourses in the United States

The discursive frames that define the climate change movement in the United States draw upon these three perspectives, and EMOs have integrated them into a number of unique approaches by combining them with their already existing discursive frames.[32] Since the US environmental movement is made up of a broad range of unique discursive communities, each community has focused on a particular interpretation of climate change that is in line with their overall discursive frame. So the diversity of approaches in the climate change movement mirrors the diversity in the overall US environmental movement. For example, the BlueGreen Alliance, an alliance of labor unions and environmental organizations, focuses on development of renewable energy legislation and increased government spending to spur job growth. Seasons' End, an alliance of sports and hunting clubs, emphasizes the threats to outdoor sporting recreation. Finally, Rising Tide, a radical direct-action environmental coalition based in deep ecology, seeks fundamental transformation of the social and economic structure of society.[33]

Within the US environmental movement, the dominant perspective is reform environmentalism. So it is not at all unexpected that the dominant discursive frame on climate change in the United States at the national level would be the form of reform environmentalism known as weak ecological modernization.[34] Perhaps the most widely recognized and influential coalition is the Climate Action Partnership, which is a coalition of large corporations and well-established environmental organizations. This coalition is the result of efforts of the Environmental Defense Fund, a high-profile environmental group that is one of the leading proponents of cap-and-trade legislation in the United States.[35] Although not a coalition, the Alliance for Climate Progress, founded by Vice President Al Gore, is also one of the high-profile EMOs advocating for a weak ecological modernization perspective. Out of the 21 coalitions, 11 adopt a weak ecological modernization perspective. Within this broad discursive frame, two coalitions, the BlueGreen Alliance and the Apollo Alliance, involve the alliance of a union with an EMO to advocate for increased government spending to create jobs in the renewable energy area. A second subgroup is made up of the Clean Energy Works and the Sustainable Energy Coalition, which focus on the development of renewable energy. A third subgroup is composed of three different religion-based coalitions that have combined concern over climate change with an ecospiritualist perspective.[36] In a similar manner, the Seasons' End coalition combined the discourse of wildlife management with a weak ecological modernization perspective regarding climate change. Finally, there are two coalitions, the Rocky Mountain Climate Organization and the Interwest Energy Alliance, that focus on western regional renewable energy development.

The second major grouping of climate change organizations centers on the discursive frame of strong ecological modernization in combination with global governance. These coalitions, primarily 350.org/1Sky, TckTckTck, and USCAN, operate across the globe in a number of countries. The focus is on the development of a mandatory global climate agreement that takes into account the needs of the developing countries and is formulated through an inclusive and democratic process. At the US national level, these coalitions focus on the implementation of mandatory limits on carbon emissions. Since these coalitions are large, they generally span a number of more particular discursive frames that are represented in their organizations in order to form a larger and more general coalition focus.

The third group of coalitions operates within the discursive frame of civic environmentalism. Two coalitions—Climate SOS and the Energy Action Coalition—advocate for large-scale economic and political change through established channels. Three other coalitions—the Mobilization for Climate Justice, Rising Tide, and Via Campesina—promote large-scale systematic change through protests, nonviolent civil disobedience, and direct action. Within the United States, Rising Tide is affiliated with the decentralized Earth First! environmental action group, which is known for direct action.

Thus with some unique US modifications, the discursive frames of the US climate change movement generally parallel the global discursive frames described by Bäckstrand and Lövbrand.[37] These discursive frames define the different perspectives that define the field of climate change politics in the United States at the national level.

Resource Mobilization Levels

A key factor in the influence of different coalitions is the level of resources mobilized to advocate for a certain approach to climate change.[38] Coalitions with greater resources can have a larger influence in the policy process. Thus an examination of the levels of resource mobilization provides an important perspective from which an entire movement's organizational structure and dynamics can be viewed. To accomplish this, two approaches were used. The first was to look at the nature and number of organizations involved in the different coalitions.[39]

There is a great deal of diversity in both the size and composition of the organizational composition of these coalitions. The largest coalition is 350.org. This is composed primarily of nongovernmental organizations (NGOs). It also has links to nine other coalitions. The structure of the two next-largest generalist coalitions, TckTckTck, and the US Climate Action Network, are quite similar. There are three coalitions that are primarily composed of for-profit corporations. These are the Interwest Energy Alliance, the Sustainable Energy Coalition, and the US Climate Action Partnership. In addition, both the Apollo Alliance and the BlueGreen Alliance are composed primarily of labor unions and NGOs.

There is a relationship between the discursive orientation of a coalition and the organizational composition. Coalitions with a discursive orientation of weak ecological modernization have the highest participation of for-profit corporations and the lowest percentage participation of NGOs. Conversely, coalitions with a civic environmentalism discursive frame have the lowest level of corporate participation and the highest involvement of NGOs. Coalitions with a strong ecological modernization discourse are similar to civic environmental organizations in that they have low corporate involvement and high NGO involvement. In addition, an examination of the absolute size of the different coalitions based on discursive frame shows an overwhelming preponderance of organizations in the larger frame of ecological modernization, with an approximate split between weak and strong ecological modernization. The number of coalitions with a civic environmentalism discourse is dwarfed by a factor of nearly 9:1 (579:65) by the number of coalitions with an ecological modernization discourse.

A second approach to understanding the climate change movement from a resource mobilization perspective is through an examination of the financial capacity each coalition and discursive frame has the potential to utilize. To examine this component, total revenue data were compiled for each coalition based on the organization's 2009 IRS Form 990 filing.[40] Again, since many organizations belong to two or more coalitions, a strict comparison of financial resources inevitably involves double counting of financial resources. However, this method does allow for a rough comparison of the relative wealth of different coalitions.

The wealthiest coalition is clearly the US Climate Action Partnership, with $2.3 billion of revenue. This total is significantly impacted by the presence of the Nature Conservancy in this coalition, which alone has $925 million in income. However, even without the participation of the Nature Conservancy, the three NGOs that participate in this coalition together have $1.379 billion in annual revenue. This is exceeded only by the US Climate Action Network, with $1.454 billion based on the participation of 61 NGOs. At the other end of the income spectrum are the coalitions Via Campesina, with only one organization that was required to file an IRS Form 990,[41] yielding under $0.5 million in income, and the coalition Rising Tide, which was composed of three national/regional organizations, none of which filed an IRS Form 990.

In general, the coalitions based on the discursive frame of weak ecological modernization are the wealthiest, with a median income of $4.3 million. However, there are nearly twice as many organizations in the coalitions with a strong ecological modernization framing. Although these organizations have fewer financial resources on average than the weak ecological modernization coalition members (median = $2.5 million), the number of organizations involved results in a rough comparability between the overall financial resources available to both strong and weak ecological modernization coalitions. Civic environmental organizations again are dwarfed by the combined financial resources of the strong and weak ecological modernization coalitions by a factor of nearly 19:1 ($4,681 million: $244 million).

Overall, the resource mobilization analysis shows the dominance of coalitions based in the discursive frame of ecological modernization. There is a rough parity in the overall number of organizations involved in either strong or weak ecological modernization, with a great level of corporate and union participation in the weak ecological modernization coalitions, and more NGO participation in the strong ecological modernization coalitions. There is also rough financial parity regarding the NGOs involved in each discursive orientation, with wealthier NGOs concentrated in weak ecological modernization coalitions. Civic environmentalism coalitions are dwarfed in both criteria of resource mobilization by ecological modernization coalitions.

Network Analysis of Social Movements

The fourth major approach to understanding the diversity within social movements is through the use of network analysis. Network analysis is predicated on the belief that social ties exert a powerful influence over organizational activities.[42] Specifically, the structural relations of interactions between concrete entities (either individuals or organizations) forms the key orienting principle. It is these relationships that affect the nature of perceptions, beliefs, and reciprocal behavioral expectations of collective behavior. By channeling resources, communications, influence, and legitimacy, social networks create shared identities and collective interests and thus promote a common cultural orientation.[43] As the exchange of information increases, organizations form stable relationships with specific partners based on their knowledge regarding the specific competencies and reliability of other members of the network. These relationships solidify over time, and future behavioral actions become regularized and routinized within the network.[44] These interactions create a dense shared cultural repository that defines a series of mutually held expectations. Thus the network of interactions creates a unique cultural identity for itself and allows the network to differentiate itself from other networks of interaction. This defines the boundaries of the network and creates a unique system of interaction that is coherent and well demarcated from other interaction patterns.[45]

The ability to control the flow of information or other critical resources within the network is a crucial component of the influence of particular members within the network. Positions are stratified according to the dependence of other positions on them for these essential resources.[46] Most interorganizational networks are made up of a core group of dominant organizations that garners the majority of resources. Members of this core group are typically densely connected with each other, but they also form ties to peripheral organizations, albeit less frequently. In other words, exchange networks tend to be hierarchical in form, where peripheral organizations control fewer resources and have fewer exchanges with the central core. Organizations in the core attempt to maintain control of the bulk of available resources by limiting their exchanges with peripheral organizations.[47] Research shows that centrality in interorganizational networks creates legitimacy, influence, and access to important resources.[48]

Thus network analysis can capture the structures that underpin individual interactions in collective behavior as well as the dynamics of social movement coalitions in the analysis of organizational coalitions. So from the network perspective, to understand the diversity of social movements requires a network analysis of the interactional patterns between the different organizations that compose the social movement.

The network structure of the 21 climate change coalitions is shown in Figure 6.2. This diagram shows shared memberships in different coalitions by organizations. As this figure illustrates, there is a great deal of diversity in the types of organizations that make up the different coalitions, as was previously described. What this analysis allows for is to ascertain the different positions that coalitions occupy within the climate change movement overall. First, the Interfaith Climate Change Network is extremely peripheral to the overall climate change movement, as it has no organizational connections with any other national coalition. Several other coalitions, including the Catholic Coalition on Climate Change, Seasons' End, and the Rocky Mountain Climate Organization have only one organizational link each that connects them to the larger movement, and so they are also peripheral coalitions.

The remaining 17 coalitions have a number of connections to other coalitions and thus are more integrated into the overall climate change movement. The Interwest Energy Alliance, the Sustainable Energy Coalition, and Interfaith Power and Light are integrated into a cluster of corporations, trade associations, and a number of other organizations. Similarly, the Apollo Alliance shares strong links with the

FIGURE 6.2   Network Structure of 21 Climate Change Coalitions

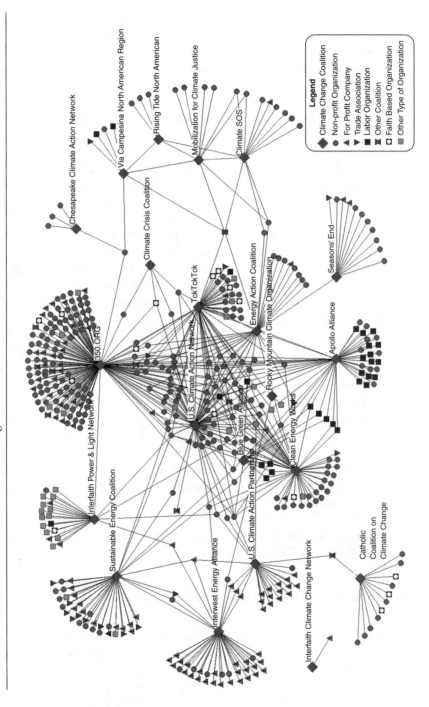

BlueGreen Alliance. The US Climate Action Partnership is somewhat marginal to the larger movement, as it shares links with only a few of the other coalitions. The remaining organizations are tightly integrated and constitute the core network components of the climate change movement.

The levels of influence and power within the overall network were examined using measures of network centrality. Centrality measures are used in social network analysis to identify the most important nodes, or those that occupy socially advantageous or influential positions in the network. Such important nodes are usually at an advantage in controlling access to information and various network resources. Additionally, these positional differences reflect status and power distinctions.[49, 50]

An examination of the centrality scores clearly shows the prominence of six coalitions that constitute the core of the US national climate change movement. The core coalitions are, first, 350.org, followed by the US Climate Action Network and TckTckTck. These three coalitions are highly central to the overall network structure. The Apollo Alliance, Clean Energy Works, and the Energy Action Coalition constitute the remaining members of the core of the network. The peripheral role of the four coalitions discussed earlier is confirmed by these data. All of the remaining coalitions are integrated into the overall network but occupy more peripheral roles than those in the core.

Moving to the organizational level, the centrality measures show a wider level of variance. An examination of the centrality scores of the NGOs in the network shows a reasonable level of correspondence between their rankings, at least as far as identifying the core organizations. Here, there are six organizations that generally score high on network centrality in all of the scores: Union of Concerned Scientists, Natural Resources Defense Council, Sierra Club, National Wildlife Federation, League of Conservation Voters, and the Rainforest Action Network.

The network analysis shows that the US national climate change movement is focused around a core of six large coalitions. While many of the other, smaller coalitions are integrated into this core, they have peripheral positions within it. Those coalitions that have high levels of for-profit involvement, including the Interwest Energy Alliance, the US Climate Action Partnership, and the Sustainable Energy Coalition, all lie outside of the core of the movement network. Finally, there are several small, very peripheral coalitions.

Power and Influence in the US Climate Change Movement

Do these institutional advantages of ecological modernization coalitions influence their capacity to influence government policy toward climate change? In his analysis, Gamson[51] measured the relative success or failure of social movements in terms of whether they gained cultural acceptance of the legitimacy of their perspectives, or if they gained new advantages regarding participation in the policy-making process. To apply these two criteria to the national climate change movement requires a focus on national-level measures of these aspects of movement success. This leads to two questions. First, which organizations have been selected as representing legitimate viewpoints that merit media coverage? Second, what organizations have been allowed to participate in government deliberations regarding national climate change policy?

To answer the first question, a list of the organizations that were cited as sources in *New York Times* stories on climate change was compiled for the time period 1980–2010. The *New York Times* is considered to be the preeminent paper in the United States and thus functions as the effective national newspaper. The legitimacy of a given perspective can thus be measured by the relative frequency of citations of organizations that represent that viewpoint. To answer the second question regarding participation in government deliberation over climate change, a count was made of the number of appearances of each organization at US congressional hearings from 1980 to 2010. Organizations are selected and invited to attend congressional hearings. So an invitation to participate can be seen as a measure of the openness to and inclusion of a particular viewpoint in government deliberations. This information is provided in Table 6.2.

This table shows a pattern similar to that of the resource mobilization analysis. While there is considerable variance between coalitions, the US Climate Action Network has both the largest number of mentions in the *New York Times* and the most appearances at congressional hearings on climate change. The coalitions with a radical civic environmentalist discourse have no mentions in the *New York Times,* nor have they ever appeared before Congress. Notably, the US Climate Action Partnership, with only four EMOs, has an enormous relative impact in the number of both mentions in the *New York Times* and appearances before Congress.

The dominance of the coalitions with an ecological modernization frame is also apparent in both *New York Times* mentions and appearances before Congress. Organizations belonging to strong ecological modernization coalitions have a slight advantage regarding mention in the *New York Times.* However, there are nearly twice as many organizations in strong ecological modernization coalitions as there are in weak ecological organization coalitions that either appeared before Congress or were mentioned in the *New York Times.* In a now-familiar pattern, the large number of organizations with a strong ecological modernization discursive frame compensated for their low per capita levels of inclusion or use as a source. Civic environmentalism is again dwarfed in both *New York Times* mentions and congressional hearing appearances.

A further empirical examination of *New York Times* mentions and congressional hearing data at the organizational level reveals the preponderant influence of a few peak organizations. There are 267 unique EMOs that were members of a climate change coalition in 2010. Out of these 267 organizations, 15 organizations

TABLE 6.2   Environmental Organizations' Inclusion in Governmental Decision Making

Organization	Number of mentions in New York Times stories on climate change	Number of congressional hearing appearances	Number of NGOs
350.org/1Sky	384	117	112
Apollo Alliance	358	107	21
BlueGreen Alliance	236	98	4
Catholic Coalition on Climate Change	0	0	6
Chesapeake Climate Action Network	13	1	4
Clean Energy Works	493	176	40
Climate Crisis Coalition	25	15	2
Climate SOS	0	0	10
Energy Action Coalition	190	21	22
Interfaith Climate Change Network	0	0	1
Interfaith Power & Light Network	152	56	5
Interwest Energy Alliance	5	4	5
Mobilization for Climate Justice	0	0	10
Rising Tide North America	0	0	5
Rocky Mountain Climate Organization	0	0	1
Seasons' End	60	51	10
Sustainable Energy Coalition	85	33	13
TckTckTck / Global Campaign for Climate Action	441	171	34
US Climate Action Network	795	298	54
US Climate Action Partnership	352	137	4
Via Campesina North American Region	0	0	3
Weak Ecological Modernization Coalitions Total	670	261	86
Strong Ecological Modernization Coalitions Total	831	301	160
Civic Environmentalism Coalitions Total	190	21	41

stand out. Each of these organizations represented 2 percent or more of the appearances at congressional hearings or mentions in the *New York Times* from 1980 to 2010. A list of these organizations and their percentage involvement in both areas is provided in Table 6.3.

Taken together, these 15 organizations represent over half of all appearances before Congress and nearly two-thirds of mentions in the *New York Times*. Some organizations, including the Environmental Defense Fund, Natural Resources Defense Council, Pew Charitable Trusts, World Wildlife Federation, and Friends of the Earth, have relatively proportional representation in both hearing appearances and *New York Times* mentions. In contrast, there are a number of organizations that have high cultural prominence and low representation before Congress or vice versa. Both the Sierra Club and Greenpeace have a high cultural profile in the *New York Times*. However, they infrequently appear before Congress. This is

TABLE 6.3   15 Major Organizations' Involvement in Government Decision Making

Organization	Congressional hearing appearances (percentage of total)	New York Times *mentions* (percentage of total)
Alliance to Save Energy	4.73	0.36
American Council for an Energy Efficient Economy	3.09	0.18
Environmental Defense Fund	7.82	13.11
Greenpeace	0.41	7.19
Natural Resources Defense Council	13.79	12.66
Pew Charitable Trusts	4.53	4.83
Resources for the Future	4.73	0.91
Sierra Club	1.85	6.83
Union of Concerned Scientists	2.88	0.36
World Resources Institute	0.41	5.19
World Wildlife Federation	2.06	5.19
Friends of the Earth	3.09	2.28
National Wildlife Federation	1.65	1.64
Center for Clean Air Policy	2.06	0.18
Nature Conservancy	2.06	1.46
Total	55.16	62.37

especially apparent with Greenpeace. The last time Greenpeace appeared before Congress in a climate change hearing was in 1990, over 20 years ago. Additionally, organizations that focus on energy efficiency, especially the Alliance to Save Energy and the American Council for an Energy Efficient Economy, are highly represented in congressional hearings but infrequently cited by the *New York Times*. Overall, what this pattern shows is the dominance of these 15 organizations in both media coverage of climate change and participation in the governmental decision-making process.

Conclusion

In this chapter, I set out to describe the growth and dimensions of the national climate change movement in the United States by focusing on three questions: How did this movement develop? What are the different viewpoints represented in this movement? What are the most influential groups? Based on this analysis, I can now offer some tentative answers to these questions. First, the movement grew out of existing environmental groups that were working on atmospheric issues in the 1970s, including acid rain and ozone depletion. As the concern over climate change grew, these organizations expanded their focus to include climate change. These original organizations have continued their involvement and now constitute a large portion of the most influential and widely recognized organizations in climate change politics. A number of new organizations and coalitions were founded in the rise of public concern that coincided with the release of Vice President Al Gore's movie *An Inconvenient Truth*,[52] and these organizations began participation in the UNFCCC meetings, and Congress increasingly called on them to testify.

The discursive viewpoint of the climate change movement in the United States at the national level is dominated by weak ecological modernization. While civic environmentalism is present, it is virtually invisible in the media and congressional testimony. This analysis provides empirical verification of the presence of what Gough and Shackley[53] label the "respectable politics of climate change." As a condition of being taken seriously in governmental discussions, organizations must adopt a discursive frame that focuses on "science and technical/policy measures and responses, and away from ethical and overtly political matters." Weak ecological modernization, since it does not challenge the existing social and political arrangements, meets this requirement. For this reason, it marginalizes the discourses of strong ecological modernization and civic environmentalism in both the media coverage and congressional testimony.

Second, there is a core of dominant institutions within the US climate change movement. The analysis shows that there is a concentrated group of 15 influential organizations, especially the Environmental Defense Fund and the Natural Resources Defense Council, that dominate both the press coverage and congressional testimony on climate change. This finding is in line with Newell's analysis.[54] He maintained that these same groups are dominant at the international level:

> The international reach of some groups derives from their access to the decision making process within powerful states. The influence of groups such as Natural Resources Defense Council (NRDC) and Environmental Defense Fund (EDF) on the Environmental Protection Agency and their ability to change the cores of votes in the US Congress have provided key leverage in achieving positive environmental outcomes in the past. At the same time, such leverage ensures that the group's voice and influence is out of all proportion to the numbers they represent.

So using Gamson's[55] criteria, only a few organizations, based in the discursive frame of weak ecological modernization, have had any success in gaining recognition of their viewpoint. However, this is only a qualified success, in that there has been no real action taken to address climate change at the federal level.

The US environmental movement is not a monolithic structure. Rather, it is composed of a number of different discursive communities. This is paralleled in the national climate change movement. Each of these communities has its own specific issue focus and network of alliances. Additionally, they have wildly varying levels of economic resources, influence, and media and press coverage. One certainty is that this movement is not static and will continue to evolve over time. Over the past century, it has developed several new discursive frames, and the structure of the movement has shifted. It will most likely continue to do so in the future. By examining this movement's structure and development, we can develop an additional dimension to our understanding of how the United States is responding to the challenge of climate change.

Suggested Readings

Bäckstrand, K., and E. Lövbrand. "Climate Governance Beyond 2012: Competing Discourses of Green Governmentality, Ecological Modernization and Civic Environmentalism." In *The Social Construction of Climate Change: Power, Knowledge, Norms, Discourses*, edited by M. E. Pettenger, 123–148. Hampshire, UK: Ashgate, 2007.

Benford, R. D., and D. A. Snow. "Framing Processes and Social Movements: An Overview and Assessment." *Annual Review of Sociology* 26 (2000): 611–639.

Brulle, Robert J., and J. Craig Jenkins. "Fixing the Bungled US Environmental Movement." *Contexts* 7, no. 2 (2008): 14–18.

Gough, C., and S. Shackley. "The Respectable Politics of Climate Change: The Epistemic Communities and NGOs." *International Affairs* 77, no. 2 (2001): 329–345.

Newell, P. "Climate for Change? Civil Society and the Politics of Global Warming." In *Global Civil Society 2005/6*, edited by M. Glasius, M. Kaldor, and H. Anheir, 90–119. Thousand Oaks, CA; Sage.

References

Adams, S., M. Jochum, and H. Kriesi. "Coalition Structures in National Policy Networks: The Domestic Context of European Politics." In *Civil Society and Governance in Europe: From National to International Linkages,* edited by W. A. Maloney and J. W. Deth, 193–217. Northamption, MA: Edward Elgar.

Bäckstrand, K., and E. Lövbrand. "Climate Governance Beyond 2012: Competing Discourses of Green Governmentality, Ecological Modernization and Civic Environmentalism." In *The Social Construction of Climate Change: Power, Knowledge, Norms, Discourses,* edited by M. E. Pettenger, 123–148. Hampshire, UK: Ashgate, 2007.

Beamish, T., and A. Luebbers. "Alliance Building Across Social Movements: Bridging Difference in a Peace and Justice Coalition." *Social Problems* 56, no. 4 (2009): 647–676.

Benford, R. D., and D. A. Snow. "Framing Processes and Social Movements: An Overview and Assessment." *Annual Review of Sociology* 26 (2000): 611–639.

Berger, G., A. Flynn, F. Hines, and R. Johns. "Ecological Modernization as a Basis for Environmental Policy: Current Environmental Discourse and Policy and the Implications on Environmental Supply Chain Management." *Innovation* 14, no. 1 (2001): 55–72.

Brulle, Robert J. *Agency, Democracy, and Nature: The U.S. Environmental Movement From a Critical Theory Perspective.* Cambridge, MA: MIT Press, 2000.

Brulle, Robert J., Jason Carmichael, and J. Craig Jenkins. "Shifting Public Opinion on Climate Change: An Empirical Assessment of Factors Influencing Concern Over Climate Change in the U.S." *Climatic Change* 114, no. 2 (2012): 169–188.

Bryner, G. "Failure and Opportunity: Environmental Groups in US Climate Change Policy." *Environmental Politics* 17, no. 2 (2008): 319–336.

Cantor, Robin, and Gary Yohe. "Economic Analysis." In *Human Choice and Climate Change,* edited by Steve Rayner and Elizabeth Malone, ch. 1, vol. 3. Columbus, OH: Battelle Press.

Christoff, P. "Ecological Modernisation, Ecological Modernities." *Environmental Politics* 5, no. 3 (1996): 476–500.

Cook, K. S., and J. M. Whitmeyer. "Two Approaches to Social Structure: Exchange Theory and Network Analysis." *Annual Review of Sociology* 18 (1992): 109–127.

Gamson, W. A. *The Strategy of Social Protest.* Homewood, IL: Dorsey, 1975.

Glover, Leigh. *Postmodern Climate Change.* New York: Routledge, 2006.

Goffman, Erving. *Frame Analysis: An Essay on the Organization of Experience.* Boston: Northeastern University Press, 1974.

Gough, C., and S. Shackley. "The Respectable Politics of Climate Change: The Epistemic Communities and NGOs." *International Affairs* 77, no. 2 (2001): 329–345.

Gulati, R., and M. Gargiulo. "Where Do Interorganizational Networks Come From?" *American Journal of Sociology* 104, no. 5 (1999): 1439–1493.

Habermas, Jürgen. *Between Facts and Norms: Contributions to a Discourse Theory of Law and Democracy.* Cambridge, MA: MIT Press, 1996.

Habermas, Jürgen. *Communication and the Evolution of Society.* Boston: Beacon, 1979.

Jenkins, J. C. "Resource Mobilization Theory and the Study of Social Movements." *Annual Review of Sociology* 9 (1983): 527–553.

Kearns, L. "Religious Climate Activism in the United States." In *Religion in Environmental and Climate Change: Suffering, Values, Lifestyles,* edited by Dieter Gerten and Sigurd Bergmann, 132–151. New York: Continuum.

Knoke, David. *Political Networks: The Structural Perspective.* Cambridge, UK, and New York: Cambridge University Press, 1990.

Knoke,, David, and S. Yang. *Social Network Analysis.* Thousand Oaks, CA: Sage, 2008.

Laclau, E., and C. Mouffe. *Hegemony and Socialist Strategy: Towards a Radical Democratic Politics.* London: Verso.

Laumann, Edward O., and David Knoke. *The Organizational State: Social Choice in National Policy Domains.* Madison: University of Wisconsin Press, 1987.

Mol, A. P. *Globalization and Environmental Reform: The Ecological Modernization of the Global Economy.* Cambridge, MA: MIT Press, 2001.

Murphy, J. "Editorial: Ecological Modernisation." *Geoforum* 31 (2000): 1–8.

Newell, P. "Civil Society, Corporate Accountability and the Politics of Climate Change." *Global Environmental Politics* 8, no. 3 (2008): 122–153.

Newell, P. "Climate for Change? Civil Society and the Politics of Global Warming." In *Global Civil Society 2005/6,* edited by M. Glasius, M. Kaldor, and H. Anheir, 90–119. Thousand Oaks, CA: Sage, 2006.

Poloni-Staudinger, L. "Why Cooperate? Cooperation Among Environmental Groups in the United Kingdom, France, and Germany." *Mobilization* 14, no. 3 (2009) 375–396.

Schlosberg, D., and S. Rinfret. "Ecological Modernization, American Style." *Environmental Politics* 17, no. 2 (2008): 254–275.

Skocpol, T. *Diminished Democracy: From Membership to Management in American Civic Life.* Norman: University of Oklahoma Press, 2003.

Snow, David A. "Framing Processes, Ideology, and Discursive Fields." In *The Blackwell Companion to Social Movements,* edited by David A. Snow, Sarah A. Soule, and Hanspeter Kriesi, 380–412. Oxford, UK: Blackwell.

Tilly, Charles. *From Mobilization to Revolution.* New York: Random House, 1978.

Wardekker, J. A, Arthur C. Petersen, and Jeroen van der Sluijs. "Ethics and Public Perception of Climate Change: Exploring the Christian Voices in the US Public Debate." *Global Environmental Change* 19, no. 4 (2009): 512–521.

York, R., E. Rosa, and T. Dietz. "Footprints on the Earth: The Environmental Consequences of Modernity." *American Sociological Review* 68, no. 2 (2003): 279–300.

Zald, Mayer N., and John D. McCarthy. *Social Movement in an Organizational Society.* New Brunswick, NJ: Transaction, 1987.

Notes

1. *Environmental Implications of the New Energy Plan*. Hearing before the Subcommittee on the Environment and the Atmosphere, Committee on Science and Technology, US House of Representatives, 95th Congress, 109 (1977).
2. Jürgen Habermas, *Between Facts and Norms: Contributions to a Discourse Theory of Law and Democracy* (Cambridge, MA: MIT Press, 1996).
3. D. Schlosberg and S. Rinfret, "Ecological Modernization, American Style," *Environmental Politics* 172 (2008): 270–271.
4. Robert J. Brulle, *Agency, Democracy, and Nature: The U.S. Environmental Movement From a Critical Theory Perspective* (Cambridge, MA: MIT Press, 2000).
5. Theda Skocpol, *Diminished Democracy: From Membership to Management in American Civic Life* (Norman: University of Oklahoma Press, 2003).
6. Ibid.
7. Habermas, *Between Facts and Norms*, 381.
8. Jürgen Habermas, *Communication and the Evolution of Society* (Boston: Beacon, 1979), 125.
9. Habermas, *Between Facts and Norms*, 381.
10. P. Newell, "Civil Society, Corporate Accountability and the Politics of Climate Change," *Global Environmental Politics* 83 (2008): 122–153.
11. Brulle, *Agency, Democracy, and Nature*.
12. T. Beamish and A. Luebbers, "Alliance Building Across Social Movements: Bridging Difference in a Peace and Justice Coalition," *Social Problems* 564 (2009): 647–676; L. Poloni-Staudinger, "Why Cooperate? Cooperation Among Environmental Groups in the United Kingdom, France, and Germany," *Mobilization* 143 (2009): 375–396; S. Adams, M. Jochum, and H. Kriesi, "Coalition Structures in National Policy Networks: The Domestic Context of European Politics," in *Civil Society and Governance in Europe: From National to International Linkages,* ed. W. A. Maloney and J. W. Deth (Northamption, MA: Edward Elgar, 2008), 193–217.
13. Erving Goffman, *Frame Analysis: An Essay on the Organization of Experience* (Boston: Northeastern University Press, 1974).
14. R. D. Benford and David A. Snow, "Framing Processes and Social Movements: An Overview and Assessment, "*Annual Review of Sociology,* 26 (2000): 611–639.
15. David A. Snow, "Framing Processes, Ideology, and Discursive Fields," in *The Blackwell Companion to Social Movements,* ed. David A. Snow, Sarah A. Soule, and Hanspeter Kriesi (Oxford, UK: Blackwell, 2004), 380–412.
16. E. Laclau and C. Mouffe, *Hegemony and Socialist Strategy: Towards a Radical Democratic Politics* (London: Verso, 1985).
17. K. Bäckstrand and E. Lövbrand, "Climate Governance Beyond 2012: Competing Discourses of Green Governmentality, Ecological Modernization and Civic Environmentalism," in *The Social Construction of Climate Change: Power, Knowledge, Norms, Discourses,* ed. E. Pettenger (Hampshire UK: Ashgate: 2007), 123–148.
18. Ibid., 126–129.
19. Leigh Glover, *Postmodern Climate Change* (New York: Routledge, 2006).

20. Bäckstrand and Lövbrand, "Climate Governance Beyond 2012," 129–131.
21. Robin Cantor and Gary Yohe, "Economic Analysis," in *Human Choice and Climate Change*, ed. Steve Rayner and Elizabeth Malone (Columbus, OH: Battelle Press, 1998), 70–71.
22. J. Murphy, "Editorial: Ecological Modernisation," *Geoforum* 31 (2000): 1–8.
23. A. P. Mol, *Globalization and Environmental Reform: The Ecological Modernization of the Global Economy* (Cambridge, MA: MIT Press, 2001), 56.
24. R. York, E Rosa, and T. Dietz, "Footprints on the Earth: The Environmental Consequences of Modernity," *American Sociological Review* 68, no. 2 (2003): 279–300.
25. P. Christoff, "Ecological Modernisation, Ecological Modernities," *Environmental Politics,* 53 (1996), 476–500.
26. Glover, *Postmodern Climate Change.*
27. G. Berger, A. Flynn, F. Hines, and R. Johns, "Ecological Modernization as a Basis for Environmental Policy: Current Environmental Discourse and Policy and the Implications on Environmental Supply Chain Management," *Innovation* 14, no. 1 (2001): 55–72.
28. Bäckstrand and Lövbrand, "Climate Governance Beyond 2012," 132.
29. Ibid.
30. Ibid.
31. Ibid., 134.
32. Brulle, *Agency, Democracy, and Nature.*
33. The self-description developed by each coalition and the dominant discursive frame of that coalition regarding climate change are available on request from the author.
34. D. Schlosberg and S. Rinfret, "Ecological Modernization, American Style," 254–275.
35. G. Bryner, "Failure and Opportunity: Environmental Groups in US Climate Change Policy," *Environmental Politics* 172 (2008): 319–336.
36. Brulle, *Agency, Democracy, and Nature.*; L. Kearns, "Religious Climate Activism in the United States," in *Religion in Environmental and Climate Change: Suffering, Values, Lifestyles,* ed. Dieter Gerten and Sigurd Bergmann (New York: Continuum, 2012), 132–151; J. A. Wardekker, Arthur C. Petersen, and Jeroen van der Sluijs, "Ethics and Public Perception of Climate Change: Exploring the Christian Voices in the US Public Debate," *Global Environmental Change* 19 (2009): 512–521.
37. Bäckstrand and Lövbrand, "Climate Governance Beyond 2012," 134.
38. C . Tilly, *From Mobilization to Revolution* (New York: Random House, 1978); W. A. Gamson, *The Strategy of Social Protest* (Homewood, IL: Dorsey, 1975); J. C. Jenkins, "Resource Mobilization Theory and the Study of Social Movements," *Annual Review of Sociology* 9 (1983), 527–553; Mayer N. Zald and John D. McCarthy, *Social Movements in an Organizational Society* (New Brunswick, NJ: Transaction, 1987).
39. Institutional composition data of coalitions available upon request from the author.
40. Financial data available upon request from the author.
41. Organizations with over $25,000 in annual revenue are required to file an IRS Information Return, IRS Form 990.
42. David Knoke, *Political Networks: The Structural Perspective* (Cambridge, UK, and New York: Cambridge University Press, 1990).

43. David Knoke and S. Yang, *Social Network Analysis* (Thousand Oaks, CA: Sage, 2008).

44. R. Gulati and M. Gargiulo, "Where Do Interorganizational Networks Come From?" *American Journal of Sociology* 104, no. 5 (1999): 1440.

45. Stephan Fuchs, *Against Essentialism: A Theory of Culture and Society* (Cambridge MA: Harvard University Press, 2001), 272–275.

46. Knoke, *Political Networks*, 9.

47. K. S. Cook and J. M. Whitmeyer, "Two Approaches to Social Structure: Exchange Theory and Network Analysis," *Annual Review of Sociology* 18 (1992), 109–127.

48. Edward O. Laumann and David Knoke, *The Organizational State: Social Choice in National Policy Domains* (Madison: University of Wisconsin Press, 1987).

49. Beamish and Luebbers, "Alliance Building."

50. Network centrality measures available from the author on request.

51. Gamson, *The Strategy of Social Protest.*

52. Robert J. Brulle, Jason Carmichael, and J. Craig Jenkins, "Shifting Public Opinion on Climate Change: An Empirical Assessment of Factors Influencing Concern Over Climate Change in the U.S.," *Climatic Change* 114, no. 2 (2012): 169–188.

53. C. Gough and S. Shackley, "The Respectable Politics of Climate Change: The Epistemic Communities and NGOs," *International Affairs* 772 (2001): 329–345, 332.

54. P. Newell, "Climate for Change? Civil Society and the Politics of Global Warming," in *Global Civil Society 2005/6,* ed. M. Glasius, M. Kaldor, and H. Anheir (Thousand Oaks, CA: Sage, 2006), 90–119, 107.

55. Gamson, *The Strategy of Social Protest.*

CHAPTER 7

Environmental Policies on the Ballot

Diana Forster and Daniel A. Smith

FROM THE PROGRESSIVE ERA ONWARD, direct participatory mechanisms such as the initiative and referendum have been leveraged by citizen groups seeking to influence state policy making. The direct democracy movement that emerged out of the western United States at the turn of the 19th century was largely driven by coalitions of special interest groups such as the Grange, temperance activists, and single taxers who wanted to take their goals directly to the people.[1] Since the adoption of direct democracy, environmentalists, just like other interest groups, have sought to achieve their protectionist objectives through ballot measures. In fact, many organizations at the forefront of climate change issues view direct democracy as a valuable political strategy, devoting substantial financial resources and volunteers to the passage of renewable energy and other environmental ballot measures. For example, in the 2012 general election alone, the Sierra Club, the Environmental Defense Fund, the William and Flora Hewlett Foundation, and numerous other interest groups marshaled their resources in support of ballot initiatives such as California's Proposition 39, the Clean Energy Jobs Act, and Michigan's Proposal 3, Michigan Energy, Michigan Jobs.

Yet we have very little basic knowledge about the extent to which direct democracy actually represents a promising venue through which environmental activists can achieve their policy goals of promoting renewable energy and arresting climate change. What is the success rate of proenvironment ballot measures across the American states? Are there notable differences across types of ballot measures, including those dealing with climate change? For example, are legislative referendums that seek to conserve land and water more successful than citizen initiatives that seek to provide alternative sources of energy? Much of the recent scholarship emphasizes determinative factors such as voter ambivalence, the campaign, and the level of campaign financing. Yet, building on the work of Magleby, we know that voter ambivalence is often steady across environmental measures regardless of the levels of spending.[2] Disproportionate amounts of money may be spent by proponents and opponents of ballot measures, and saliency of environmental measures tends to be low relative to other issues on the ballot. Setting these factors aside, the structural environment is also likely to have an impact. Ballot measures are not all equal in terms of how they qualify for the ballot and when they appear before voters. Initiatives and legislative referendums are unlikely to be equally effective when they are used in environment-related decision making. Other factors, such as the type of election in which a proenvironment measure appears on the ballot and the total number of measures on the ballot, are also likely to be important. As our findings show, these institutional factors can play a decisive role in the success of proenvironment ballot measures.

This chapter offers a multimethod study exploring the usage patterns and passage rates of proenvironment ballot measures at the state level from the Progressive Era to the present. We draw on a complete dataset of 349 diverse proenvironment ballot measures drawn from 35 states from the period of 1908 to 2010 to empirically examine the correlates of success for proenvironment initiatives and referred measures and assess how they compare to the traditional predictors of passage rates. After a review of recent scholarship on public opinion, the environment, and direct democracy, we further elucidate the tripwires to passage of proenvironment ballot measures by presenting a series of contrasting case studies. We conclude by discussing the future prospects of environmental measures at the ballot box, particularly those dealing expressly with climate change, and the ongoing incentives for environmental activists to put their reforms directly before voters.

Public Opinion on the Environment

As Dennis Chong's chapter in this volume makes clear, environmental issues occupy a uniquely ambiguous place in public opinion. With the exception of major natural disasters, environmental policy is rarely a top-of-the-head issue for voters.[3] When measured with such mechanisms as the traditional "most important problem" survey probe, environmental issues are generally deemed to be low salience.[4] Although voters have deemed energy policy an important concern in the 2008 and 2012 elections, their anxieties typically revolve around the immediate challenge of rising gas prices rather than the broader implications of climate change.[5] Additionally, survey research has found that public responses to questions around support for protecting the environment are highly sensitive to framing effects and changes in word choice, and that they vary across time and in accordance with economic conditions.[6]

Early research held that environmental policy represented a form of "consensus politics" within the United States, with support for environmental protectionism in Congress cutting across both partisan and regional cleavages.[7] However, contemporary scholars have found a strong partisan cleavage on environmental policy in that Republicans are significantly less likely to support protectionist regulations than

Democrats, and there are clear differences in environmental policy among partisans and ideologues from the elite level all the way down to the level of mass opinion.[8]

Despite, or perhaps because of, this pattern of party ownership, scholars disagree over the broader importance of environmental issues to national politics. Guber uses American National Election Study (ANES) data to raise doubts about the ability of environmental policy to exert influence over national elections.[9] Her study concludes that environmental issues are too low salience to draw significant numbers of Republican issue voters toward the Democratic Party. However, in keeping with previous findings from survey research on environmental issues, Davis and Wurth use a different measure of concern for the environment to argue that Guber's research design severely underestimates the salience of environmental issues; they find that support for the environment was a significant predictor of a vote for Clinton over Dole in the 1996 presidential election.[10]

Setting aside questions of frame and salience, citizens—even Republicans— are generally supportive of protecting the environment in the abstract. In the United States as well as western Europe, environmentalism enjoys a higher level of public approval in surveys than the goals of any other contemporary social movement.[11] National polls show that a majority of respondents believe that government spending on the environment is too low and that government regulations around environmental protection need to be increased.[12] More so than on other issues, Americans are highly skeptical that business interests will protect the environment on their own; hence, most believe that government intervention is necessary in this area. Even more telling, majorities of survey respondents in the US say that they support environmental protection even at the expense of economic growth and that they are willing to pay higher prices in order for this to happen.[13] However, even climate change, one of the most highly publicized environmental issues in American history, remains low salience for US voters.[14]

But are voters willing to put their money where their mouths are when given a chance as citizen lawmakers to vote on these issues? Do voters support proenvironment measures at the ballot box, or do they reject the financial costs and the legal constraints these measures often place on businesses? In short, what explains the success (or failure) of proenvironment ballot measures? Are the generally accepted predictors of passage rates for ballot measures the same as those for proactive environmental protectionist measures?

Why Proenvironment Ballot Measures Succeed (and Fail)

Studies of ballot measures dealing broadly with the environment have highlighted four aspects of the campaigns: the level of information available to voters, campaign financing, ballot language and framing effects, and the ambiguity inherent in public perceptions toward the environment.

First, with respect to the level of information on environmental ballot measures that is available to voters, although these measures tend to have generally high levels of support, it is often inconsistent. In general, voters are known to have low levels of information about ballot measures.[15] Environmental measures are no different in this regard.[16] When environmental issues are low salience, it may be a challenge for voters to render informed decisions about them, regardless of what their "true" levels of support are. Additionally, ballot measures typically have highly technical and even convoluted language, which poses another challenge to voter decision making on environmental policy.[17]

At the same time, previous studies have found that voters are able to employ shortcuts and heuristics to make informed decisions on ballot measures on which they have very little information.[18] For example, partisanship has been shown to be a strong heuristic even in the absence of partisan cues.[19] Given the extent to which support for environmentalism has been found to align with partisanship and ideology in the United States, it is likely that partisanship is a reliable heuristic for voters on environmental measures. However, as we discuss below, this process may be contingent upon framing and elite endorsements.[20]

The low levels of information available on environmental ballot measures in advance of elections and the malleability of public opinion on them suggest that, more so than in many other issues where public opinion is thoroughly entrenched, support for ballot measures addressing environmental interests may be highly sensitive to campaign effects.[21] To use Magleby's language, environmental measures occupy an area of "uncertain opinions," where voter opinion may fluctuate significantly or reverse entirely over the course of the campaign.[22] In the case of environmental measures, as with many other ballot measure campaigns, the side that defines the issue for the public tends to be the side that wins the election.[23]

Previous studies concerning information levels and environmental ballot measures have almost exclusively focused their analyses on a small number of initiatives and thus have borne out few consistent findings. Lake examines environmental measures that were on the ballot in California in the 1970s.[24] Looking at passage rates, she finds that environmental bonds had a slightly higher passage rate than nonenvironmental bonds, although most environmental spending measures on the ballot during this period were successful. Although initiatives were less likely to pass, their passage rates during this period were no lower than those of nonenvironmental initiatives. Her data suggest that California's environmental measures have a broad base of popular support, crossing socioeconomic, ethnic, generational, and partisan lines.

When it comes to campaign financing of ballot measures, several studies have pointed out the importance of money to the support or failure of progressive environmental policies. Campaign spending has been a consistently strong predictor of success at the ballot box, particularly in cases where the opposition involves well-funded business interests, as it does in most environmental campaigns. To this end, Gerlak and Natali look at campaign spending data on twelve 1992 environmental measures, three-quarters of which ultimately failed.[25] They find that the proenvironment side was outspent on all nine of the measures that failed, leading them to conclude that environmental groups suffer at the ballot box due to being consistently and significantly outspent by their opponents.

More generally, research has suggested that campaign spending by opponents has a greater impact than spending in favor of a ballot measure.[26] There are few regulations on the financing of ballot initiatives across the states, and no state limits on campaign contributions or expenditures.[27] The courts have struck down all requirements around campaign spending with the exception of disclosure, which donors have been able to circumvent by creating "veiled political actors," able to conceal campaign contributions by using patriotic or populist-sounding names to disguise their supporters.[28] For example, Garrett and Smith cite the case of Gerald Meral, the president of the nonprofit Planning and Conservation League Foundation (PCL/F) and former executive director of that organization's lobbying arm, the Planning and Conservation League (PCL). Meral exploited the loopholes in California's campaign finance laws in order to create a number of shell organizations that contributed to the passage of California's Proposition 50 in 2002, a bond initiative that sought to improve water quality and protect the coastal wetlands area.

While the high levels of campaign spending and occasionally opaque and even deceptive financing methods are clearly a challenge to the progressive reputation of direct democracy, there is no consensus among scholars on the basic question of whether spending can really impact campaign outcomes. Even research projects that move beyond case study methodology and attempt to measure campaign spending must contend with the endogeneity of spending on ballot measure campaigns. To this end, de Figueiredo, Ji, and Kousser contend that the finding by Magleby and others that spending against a measure is more influential than spending in support of a measure results from the fact that initiative proponents spend more when their issue has a higher probability of failure.[29] Examining the financing of California ballot propositions between 1976 and 2004, they find that spending on both sides of an initiative influence the probability of passage in parallel ways. Similarly, Stratmann attempts to circumvent the endogeneity issue by using county-level data on television advertising for and against California ballot measures in the early 2000s.[30] Like de Figueiredo and his coauthors, Stratmann finds that campaign spending matters on both sides of the campaign, with the likelihood of passage increasing with the level of spending by proponents on, and decreasing in tandem with, opposition spending on ads. In fact, he finds that spending by the supporting side has a greater marginal impact than spending by the opposition in some of his model specifications.

On a third front, several authors have suggested that focusing on campaign spending as an explanation for the passage rates of proenvironment measures is overly simplistic. Instead, they recommend taking a broader view of the campaign effects of ballot language and issue framing. In an early study, Lutrin and Settle contrast experiences with two initiatives that were on the ballot in California in 1972, Proposition 9, the Clean Air Initiative, and Proposition 20, the Coastal Zone Conservation Act.[31] They argue that the latter passed while the former failed, because the latter had a broad-based coalition supporting it and lacked the strong, well-funded opposition campaign that defeated the Clean Air Initiative. Lutrin and Settle also point out that the Coastal Zone Conservation Act was written in positive, clear language, while the title and ballot language of the Clean Air Initiative were more convoluted.

Using a research design similar to that of Lutrin and Settle, Guber contrasts two initiatives, one that passed and one that failed.[32] California's 1986 toxics initiative, Proposition 65, passed, benefiting as the Coastal Zone Conservation Act did from appealing language that captured uninformed voters at the ballot box. Meanwhile, Massachusetts's 1992 recycling initiative, Question 3, failed, because the well-funded opposition was able to frame the debate with an emphasis on government waste and bureaucracy. Thus, reframing the debate was the key to the success of Question 3's opposition campaign.

Finally, in keeping with the findings of previous studies of survey responses to questions on the environment, Chong and Wolinsky-Nahmias argue that voter opinion on growth and conservation campaigns can be marked by uncertainty and ambivalence.[33] Voters "generally favor protection of the environment and the public purchase of open space, but are also concerned about sustaining economic growth and are reluctant to pay higher taxes for conservation programs."[34] While the public generally supports environmental protections in the abstract, at the ballot box they are often pragmatically concerned about their potential effects on economic interests.

It is certainly possible that voter ambivalence on a ballot measure is affected by campaign financing. As Gerlak and Natali show, corporate interests that typically oppose, for example, open space initiatives tend to be better funded than the proponents of these issues are.[35] Chong and Wolinsky-Nahmias present a case study of Arizona's Proposition 202, the Citizen's Growth Management Initiative (CGMI), which was sponsored by the Sierra Club. On the ballot in 2000, CGMI sought to combat urban sprawl by promoting growth management plans for cities and large counties, increasing infrastructure costs for builders and empowering citizens to vote on general development plans. Despite high levels of voter support for the initiative early on, it was ultimately defeated by a $5 million-plus effort by Arizonans for Responsible Planning, a coalition of developers and a subsidiary of the state homebuilders association. The strong opposition campaign effectively emphasized the financial costs of the measure. Ballot measures that seek to protect the environment are most likely to pass when they fall under the radar and in the absence of meaningful opposition. When proponents "are able to conduct essentially one-sided campaigns in which they remain free to frame the benefits of the policies they promote," Chong and Wolinsky-Nahmias conclude, voter ambivalence succumbs to support for environmental protection.[36]

Bornstein and Lanz's study suggests a similar pattern of ambivalence with regard to progressive environmental policies, especially ones that have a direct impact on the voter's pocketbook.[37] They examine three ballot measures that were under consideration in Switzerland in 2000, each of which proposed taxes on fossil fuels, and each of which failed. They argue that although the financial effects negatively influenced voters' probability of supporting the initiatives, voters also experienced social pressures to support the initiatives, which were widely perceived as benefiting the larger public good. Additionally, they find that while economic concerns influenced support, their impact was blunted by controls for ideology. In other words, liberal voters were still more inclined to support the taxes, even when their practical financial considerations were taken into account.

With the exception of Lake's examination of municipal bonds, all of the previous studies of environmental issues and direct democracy have focused exclusively on initiatives.[38] This is dangerously limiting, as initiatives face high hurdles at the ballot box and are more likely to face strong opposition campaigns, such as those emphasized by Lutrin and Settle, Guber, and Chong and Wolinsky-Nahmias. Bonds and measures that are referred by the legislature consistently have higher passage rates than citizen initiatives, as they tend to be both less publicized and less innately controversial.[39] Thus, it is likely that the findings around environmental initiative campaigns do not hold for legislative referrals. The pattern of passage rates for referred measures, for which the threat of a well-funded opposition is much less dire, is likely much higher than the pattern for initiatives.

In sum, previous studies of the success or failure of environmental ballot measures have emphasized the information available to voters, the impact of money, framing effects of the ballot language, and the inherent ambiguity of public perceptions toward environmental issues. However, their generalizability is limited by their use of case study methodology and their focus mostly on ballot initiatives. They have also failed to examine some of the traditional predictors of success at the ballot box, such as ballot crowding and the type of election. If awareness is higher in midterm elections than in general elections, as Bowler and Donovan and Nicholson have found more generally, then proenvironment measures on the ballot in midterm elections may be less likely to fly under the radar and pass without controversy.[40] Voter fatigue caused by the presence of a large number of measures on statewide ballots may also decrease the probability of passage.[41]

Patterns of Proenvironment Ballot Measure Success

In order to evaluate previous research on environmental ballot measures, we created an original dataset of 349 proactive environmental initiatives and legislative referendums that were on the ballot in 35 states between 1908 and 2010. We excluded from the data set any ballot measures that sought to alter policy in any way that would negatively impact the environment—for example, those that supported economic growth at the expense of land conservation or protected the rights of individuals to hunt and fish. While we acknowledge the importance of campaign finance and framing in determining the success or failure of ballot initiatives, we focused our analysis instead on institutional factors. From a macro-level perspective, we sought to explore the impact of traditional predictors of success at the ballot box for initiatives and referred measures that seek to protect the environment. The influence of these institutional factors is a critical question for proponents of environmental reform who are seeking to leverage direct democratic policy-making mechanisms.

We begin our discussion of our work here by providing a brief historical overview of the success of all proenvironment measures on the ballot over the past century. These historical trends demonstrate that the appearance of environmental measures at the ballot box is nothing new, even though the focus of these policies has changed through the years.

Over the past century, very few statewide initiatives and referendums on the ballot in US elections have dealt expressly with climate change issues. However, measures dealing with the environment more broadly have been placed before voters since the earliest days of direct democracy. Furthermore, since the 1980s, measures promoting renewable energy in an effort to halt climate change have become increasingly prominent on state ballots.

The ballot measures in our data set encompass six broad categories: oil and gas, wildlife, nuclear power, land, air and water, and renewable energy. Because this data set is limited to proactive environmental measures, the majority of the oil and gas measures involve proposals to impose a gasoline tax. The wildlife category encompasses measures that seek to protect animals, including but not limited to fish and game. The nuclear power measures are primarily drawn from the 1970s and 1980s and include the highly publicized nuclear freeze initiatives—many of which were purely symbolic—that sought to halt the weapons race against the Soviet Union, as well as other measures regulating nuclear waste. The land category includes measures that strive to prevent land speculation and promote conservationist efforts as well as general recycling measures to eliminate landfill waste. Similarly, the air and water category covers measures that seek to protect clean air and water and includes measures such as bonds for water pollution control. Finally, the renewable energy measures seek to promote energy efficiency and alternative energy sources over fossil fuels. Figure 7.1 summarizes these six types of measures.

FIGURE 7.1   Types of Environmental Ballot Measures, 1908–2010

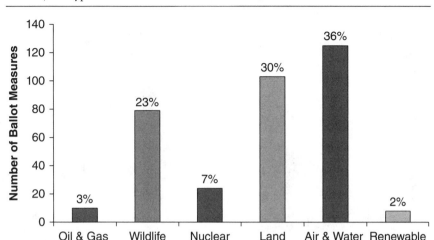

The bulk of environmental ballot measures focus on protecting wildlife, land, air, and water. These measures have a strong historical precedent, appearing even in the earliest years of direct democracy's use. For example, in 1908, Oregonians famously voted on salmon fishing laws proposed by alternate fishery operators— Measure 10, sponsored by the Fishwheel Operators, and Measure 17, proposed by the Gillnet Operators. Voters approved both measures. By contrast, measures promoting renewable sources of energy did not begin to appear until the 1970s. Notably, almost no measures, even in recent elections, directly address climate change by name. Perhaps as a strategic move to appeal to voters by employing bland, noncontroversial language, most of the measures that directly seek to stave off climate change actually invoke the language of renewable energy.

Concomitant with the growing saliency of environmental politics in the later part of the 20th century, there was a significant increase in environmental ballot measures overall in the 1970s and into the late 1980s. This growth in proenvironment ballot measures parallels the increase in the use of direct democracy in the United States more generally, with the greatest surge in the 1990s.

Passage rates vary significantly across the six categories of proenvironment ballot measures, as Figure 7.2 shows. Measures that seek to protect air and water are the most likely to pass, at 77 percent, with measures related to land conservation close behind, at 70 percent. The highly salient nuclear power initiatives have the lowest passage rate, only 38 percent. The passage rate for measures that seek to regulate oil and gas is not much higher, at 40 percent. However, in keeping with Lake's findings on California environmental ballot measures in the 1970s, the

FIGURE 7.2   Proenvironment Ballot Measures, 1908–2010 Passage Rates by Category

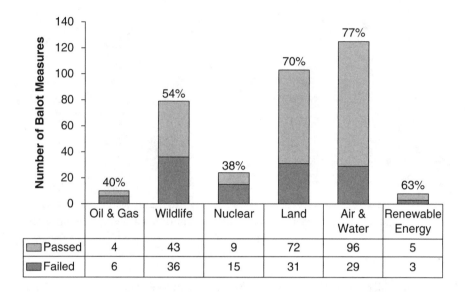

	Oil & Gas	Wildlife	Nuclear	Land	Air & Water	Renewable Energy
Passed	4	43	9	72	96	5
Failed	6	36	15	31	29	3

passage rates overall are comparable to, if not higher than, those for nonenvironmental measures.[42]

No previous study has examined the passage rates of environmental protectionist initiatives versus legislative referendums, either historically or in contemporary elections. As previously stated, in addition to having a broader base of support than initiatives, legislative referrals may also be more likely to pass, because they often have bipartisan support in the state legislature and are less likely to face strong, well-funded opposition campaigns.

As expected, we find a significant difference in the passage rates of initiatives and referendums. The 349 ballot measures in the data set include 140 citizen initiatives and 209 legislative referendums. Overall, the initiatives had only a 40 percent passage rate, while referendums had a passage rate of more than twice that, nearly 83 percent.

Isolating the Causal Factors of Proenvironmental Ballot Measure Success

In order to further assess the macro-level factors for the passage rates of proenvironment ballot measures, we present the results of a series of three logistic regressions employing commonly used predictors of success at the ballot box, with marginal effects and predicted probabilities utilized for ease of interpretation. These models include the full data set of 349 initiatives and referendums. We exclude popular referendums—citizen petitions challenging (or vetoing) laws passed by state legislatures—due to their small number and exceptional circumstances. The level of analysis is a ballot measure in a given state in a given year. The dependent variable in each model is passage, and it is coded 1 if the proenvironment ballot measure passed and 0 if it failed. In light of the vastly different passage rates for initiatives and legislative referrals, we include a dummy variable for initiatives. We also control for statutory ballot measures in contrast to constitutional amendments. In addition, previous research on the success of ballot measures has found substantial differences across general, primary, and special elections.[43] Bowler and Donovan find higher passage rates in midterm elections, likely because turnout is higher in general elections. To explore whether this finding holds for proenvironment initiatives, we control for primary and special elections, with general elections serving as the excluded category.

Bowler and Donovan also point out the challenges related to crowded ballots, which can lead to higher levels of voters rolling-off down ballot due to voter fatigue, or voters simply following the prescription, "When in doubt, vote no."[44] Thus, we control for the total number of measures on the ballot. As a further control for the general public mood toward initiatives and referendums, we include a variable representing the percentage of total measures on the ballot that pass (excluding the ballot measure under consideration within the model).

Given the substantial historical variation for proenvironment measures in this data set and the consistent finding that environmental policy has been increasing in saliency since the 1970s, we include a simple count term that captures the elapsed time since 2010. The increased frequency of environmental measures on the ballot may be connected to voter fatigue, as voters are faced with even more environmental issues, which might be making it more difficult for measures to pass today than a decade or more ago. On the other hand, the increased saliency of environmental policy may make it easier for contemporary measures to pass. The values on this variable range from 0, for measures that appeared on the ballot in 1908, to 102, for those that were under consideration in 2010. Finally, we include dummy variables for five of the six substantive categories of environmental measures in the data set, with land use as the excluded category. All three models are clustered by state and employ robust standard errors.

The results of the base model, Model A, are presented in the first column of Table 7.1. As with the bivariate analysis, proenvironment initiatives are significantly less likely to pass than legislative referendums. The marginal effects show that an initiative has a 40 percent lower probability of passage, holding other variables at their mean values.

The measure of year count from 1908 to the present is positive and significant, suggesting that over the last 100 years, voters are more likely to support proenvironment measures, be they initiatives or referendums. The marginal

TABLE 7.1 Causal Factors for Proenvironment Ballot Measure Success

	Model A			Model B			Model C		
	Coefficient	p-value	MFX	Coefficient	p-value	MFX	Coefficient	p-value	MFX
Statutory amendment	-.300	.386	-.067	-.286	.471	-.063	-.267	.551	-.058
Primary election	.092	.855	.020	.095	.793	.020	.016	.972	.003
Special election	-.499	.467	-.115	-.610	.378	-.141	-.390	.574	-.087
Initiative	-1.857	.000	-.403	-1.680	.000	-.363	-1.709	.000	-.365
Year count	.011	.034	.002	.004	.496	.001	.006	.326	.001
Number of measures on the ballot				-.001	.922	-.000	-.008	.505	-.002
Passage rate				1.957	.000	.415	2.143	.000	.448
Air and water							.748	.015	.149
Nuclear power							-.099	.900	-.021
Wildlife							.755	.127	.144
Renewable energy							.476	.575	.090
Oil and gas	.688	.187		.026	.963		-.477	.566	-.108
Constant							-.607	.399	
Wald χ^2	101.58	.000		150.39	.000		349.03	.000	
Pseudo R^2	.1632			.2042			.2230		
N	350			349			349		

Notes: Unstandardized logistic regression coefficients reported. (Robust standard errors are excluded.) General elections, legislative referrals, and land measures are omitted as excluded categories.

effects suggest that with every 10 years, environmental measures become approximately two percentage points more likely to pass, all else equal. This is likely the result of the increased salience of environmental policy and the increased levels of support for environmental protectionism among modern voters, providing evidence of strengthening postmaterialist values. The remaining variables are all insignificant. Contrary to Bowler and Donovan and other authors who have emphasized the lower passage rates in general elections, the results suggest that this pattern does not hold for proenvironment measures.[45] Furthermore, we find no evidence that statutory measures are more or less likely to pass than constitutional amendments.

Model B adds two additional variables, the total number of measures on the ballot in a particular election and the overall passage rate of the measures on the ballot in a given state in a given year, excluding the proenvironment measure under consideration. Initiatives continue to be significantly less likely to pass than legislative referrals, although the marginal effect drops slightly, from 40 percent to 36 percent. However, controlling for the total number of measures on the ballot and the overall passage rate eliminates the significance of the year count variable, suggesting that contemporary environmental measures are more likely to pass than their earlier counterparts, simply because passage rates in general have increased over time. The variable is positive, and the marginal effects show that for every percentage point increase in the passage rate, the probability of an individual proenvironment measure passing increases by 41 percent. The number of measures on the ballot is insignificant; there is no evidence that ballot fatigue, induced by having more measures on the ballot, has any influence on a proenvironment measure's chances of passage. The indicator variables for statutory amendments, primary elections, and special elections remain insignificant.

Finally, Model C adds in five of the six category variables, with measures dealing with land use again acting as the excluded category. Only measures dealing with air and water have significantly different passage rates from those dealing with land policy. The marginal effects show that measures addressing air and water quality are approximately 15 percentage points more likely to pass than those dealing with land politics. Returning to Chong and Wolinsky-Nahmias's study, it is possible that measures dealing with air and water quality are less likely to encounter the same fierce opposition by corporate interests than measures dealing with land conservation often do.[46]

Notably, renewable energy measures that deal most directly with climate change do not have significantly different passage rates from measures dealing with land policy. The coefficients for nuclear power and oil and gas regulation are negative but insignificant; the coefficients for wildlife and renewable energy are positive but also insignificant. The null effect on wildlife measures is somewhat surprising, in light of the great success these measures had in the 1990s with the Humane

Society. Thus, while there does appear to be some variation across the substantive categories of environmental measures, the variation is somewhat less substantial when controlling for other factors that predict success at the ballot box.

Controlling for the measure's specific category does not significantly change the findings of Model B. Initiatives remain significantly less likely to pass than legislative referrals, and the overall passage rate continues to be a strong and positive predictor of a measure's success. The coefficients for statutory amendments, primary elections, special elections, the year count, and the number of measures on the ballot remain insignificant. The results of this model again confirm that, in the case of proenvironment initiatives and referrals, the type of election has no significant impact on the measure's probability of passage.

The predicted probabilities presented in Figure 7.3 further elucidate the differences between initiatives and referred measures. Initiatives dealing with air and water have a 57 percent probability of passing, while legislative referrals, all else equal, have an 88 percent chance. Similarly, a land initiative has only a 39 percent chance

FIGURE 7.3   Predicted Passage Rates of Proenvironment Ballot Initiatives and Referred Measures

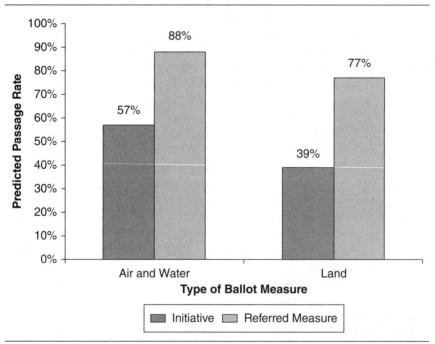

Note: Predicted probabilities are based on Table 7.1, Model C, and are calculated holding other variables at their median values.

of passing, while a referred measure dealing with land has a 77 percent likelihood of passing, ceteris paribus.

Environmental Initiatives and Referendums: Four Case Studies

When it comes to the success or failure of an environmental ballot measure, the unique differences between an initiative campaign and a legislative referendum campaign (or lack thereof), as well as the role of the general election, can be illustrated with a series of four case studies of renewable energy measures geared toward staving off the effects of climate change. All four of these measures appeared on statewide ballots within the last decade. Interestingly, despite the clear implications of these measures, the unequivocal language of "climate change" rarely, if ever, appears in either the ballot language or the campaign materials of statewide proenvironment measures. Rather, the ballot language employs the rhetoric of "renewable energy" and focuses on the immediate goals that the measure seeks to achieve. This avoidance likely results from the unwillingness of ballot measure proponents to invoke a term that may turn off conservative voters.[47]

Colorado's Renewable Energy Act

Colorado's Amendment 37, the Renewable Energy Act, was an initiated state statute that appeared on the 2004 general election ballot. It required Colorado's largest utility retailers to obtain 3 percent of their electricity from renewable energy resources by 2007, and 10 percent by 2015. The amendment further mandated that 4 percent of the renewable energy come from solar resources. The complicated ballot language read as follows:

> An amendment to the Colorado revised statutes concerning renewable energy standards for large providers of retail electric service, and, in connection therewith, defining eligible renewable energy resources to include solar, wind, geothermal, biomass, small hydroelectricity, and hydrogen fuel cells; requiring that a percentage of retail electricity sales be derived from renewable sources, beginning with 3% in the year 2007 and increasing to 10% by 2015; requiring utilities to offer customers a rebate of $2.00 per watt and other incentives for solar electric generation; providing incentives for utilities to invest in renewable energy resources that provide net economic benefits to customers; limiting the retail rate impact of renewable energy resources to 50 cents per month for residential customers; requiring public utilities commission rules to establish major aspects of the measure; prohibiting utilities from using condemnation or eminent domain to acquire land for generating facilities used to meet the standards; requiring utilities with requirements contracts to address shortfalls from the

standards; and specifying election procedures by which the customers of a utility may opt out of the requirements of this amendment.

Note that the ballot language for this measure, as with the other case studies we present, makes no direct reference to climate change whatsoever, despite the clear implications of the policies it promotes.

The *Blue Book*, which the state of Colorado publishes to provide voters with information to enable them to make informed decisions about the variety of initiatives and referendums on the ballot, summarized the arguments in favor of the measure as emphasizing the economic and environmental payoffs of renewable energy. Unlike many measures that struggled to overcome voters' ambivalence about spending hikes on the environment, Amendment 37 was presented as a "no cost" measure for citizens.[48] Meanwhile, the opposition emphasized the cost to utility companies.

Spending for and against the amendment was approximately equal. Coloradans for Clean Energy spent a total of $1,446,578 in support of the measure, and the measure received additional support from the national organization the Sustainable Energy Coalition. The two organizations lobbying against it, Citizens for Sensible Energy Choices, and the Intermountain Rural Electric Association, spent over $1.3 million combined.[49] The measure also faced a strong campaign from Xcel Energy, one of the state's leading utility companies. In the end, the Renewable Energy Act passed with 54 percent of the vote, making it the first renewable energy standard to pass through a direct democratic mechanism rather than legislative action.

California's Alternative Fuels Initiative

California's Proposition 10, the California Alternative Fuels Initiative, was on the statewide ballot in the November 4, 2008, general election. It was also known as "the Pickens Plan," after Texas oil financier T. Boone Pickens, who was instrumental in placing the measure on the ballot and came to be identified with it through the opposition campaign. The proposition was titled, "Alternative Fuel Vehicles and Renewable Energy. Bonds. Initiative Statute," and it sought to provide a total of $5 billion in general obligation bonds for a variety of renewable/clean energy purposes. If approved, the initiative would have given rebates of up to $50,000 to Californians who purchased cars that ran on natural gas or other alternative fuels, and it would have also provided funding for research into alternative energy. The summary, which appeared in the official voter guide, stated that the initiative

- Provides $3.425 billion to help consumers and others purchase certain high fuel economy or alternative fuel vehicles, including natural gas vehicles, and to fund research into alternative fuel technology.

- Provides $1.25 billion for research, development, and production of renewable energy technology, primarily solar energy with additional funding for other forms of renewable energy; incentives for purchasing solar and renewable energy technology.
- Provides grants to cities for renewable energy projects and to colleges for training in renewable and energy efficiency technologies.

The guide further stated, "Total funding provided is $5 billion from general obligation bonds."

In contrast to the ballot measures examined by Chong and Wolinsky-Nahmias, Gerlak and Natali, Guber, and others who have emphasized the importance of campaign spending in defeating environmental measures, the supporters of Proposition 10 drastically outspent the opponents of the initiative.[50] In 2008, the campaign committee that advocated for Prop 10—"Californians for Energy Independence—Yes on Prop 10, A Coalition of Renewable Energy and Alternative Fuel Companies"—received more than $22 million in contributions. Pickens's natural gas fueling company alone, Clean Energy Fuels, gave close to $19 million. By contrast, the group "No on Proposition 10, Californians Against the $10 Billion Lemon" received only $173,218 in donations and spent less than one one-hundredth as much as the initiative's supporters.[51]

The opposition included a broad coalition of consumer federations, educators, nurses, and state and municipal employees. Their campaign emphasized the high cost of the initiative to the state, which was already in the midst of an economic recession, and played up the involvement of Pickens in the campaign.[52] The opposition employed the slogan, "It's about GREED, not Green." The No on 10 Campaign claimed, "It doesn't clean the air but it does line the pockets of the billionaire who paid for it," alluding to Pickens's involvement.[53] Speaking for the No on Prop 10 campaign, Executive Director of the Consumer Federation of California Richard Holober typified the framing efforts of the opposition, saying, "Proposition 10 is the ultimate example of a wealthy special interest abusing the ballot initiative process to enrich itself."[54] In a sense, the opponents were able to leverage the ambiguity surrounding the measure's intended beneficiaries to frame it as a corporate plot rather than an environmental good. The majority of the major newspapers in the state also came out against the initiative. On Election Day, voters defeated the measure, with 59.5 percent opposing the initiative. Even with an enormous financial advantage, the initiative inspired a robust opposition campaign, which ultimately succeeded in framing the issue as promoting corporate rather than citizen interests.

Maine's Energy Efficiency Bonds Issue

Maine's Question 2, the Energy Efficiency Bonds Issue, was placed on the June 8, 2010, primary ballot by the state legislature. Similar to California's Prop 10, it called for bond spending toward energy efficiency at state colleges and elsewhere. The ballot language read simply,

> Do you favor a $26,500,000 bond issue that will create jobs through investment in an off-shore wind energy demonstration site and related manufacturing to advance Maine's energy independence from imported foreign oil, that will leverage $24,500,000 in federal and other funds and for energy improvements at campuses of the University of Maine System, Maine Community College System and Maine Maritime Academy in order to make facilities more efficient and less costly to operate?

A referred measure, the legislation was sponsored by a number of Democratic legislators, including Rep. Hannah Pingree, Rep. Seth Berry, Rep. Emily Ann Cain, Rep. John Piotti, Sen. Bill Diamond, and Senate President Elizabeth Mitchell. In order to place the referendum on the ballot, Maine required a two-thirds vote in both the House and the Senate. Question 2 received endorsements from a number of high-profile political leaders, including Democratic Governor John Baldacci, the Bangor Region Chamber of Commerce, and the Portland Chamber of Commerce, as well as most of the state's major newspapers. Its primary opponent was the Maine Heritage Policy Center, a conservative think tank, which argued that the state was in too much debt financially to spend the money.

Framed as a pro–renewable energy measure, Question 2 passed easily, garnering 59.2 percent of the vote. Neither the proponents nor the opposition mounted a vigorous campaign in the state, suggesting that most voters encountered the measure for the first time in the voting booth. The measure also benefited from a generally positive mood toward ballot measures that year; all five of the measures on the ballot in that primary election passed.

Michigan's Renewable Energy Amendment

Finally, Michigan's Proposal 3, a Renewable Energy Amendment along the lines of Colorado's successful Amendment 37, was an initiated constitutional amendment that appeared on the 2012 general election ballot. It mandated that by 2025, one-quarter of the state's electricity must come from renewable resources. The ballot language read as follows:

This proposal would:

- Require electric utilities to provide at least 25% of their annual retail sales of electricity from renewable energy sources, which are wind, solar, biomass, and hydropower, by 2025.

- Limit to not more than 1% per year electric utility rate increases charged to consumers only to achieve compliance with the renewable energy standard.
- Allow annual extensions of the deadline to meet the 25% standard in order to prevent rate increases over the 1% limit.
- Require the legislature to enact additional laws to encourage the use of Michigan made equipment and employment of Michigan residents.

Proposal 3 was filed by Michigan Energy, Michigan Jobs, a bipartisan coalition of business, environmental groups, labor organizations, and others that advocates for renewable energy and energy efficiency with the goal of job creation. Additional support was provided by the Sierra Club, the Michigan Nurses Association, the Michigan Environmental Council, the United Auto Workers, and the Sterling Corporation. The supporters projected that if the measure was approved, it would provide more than 40,000 jobs and attract $10 billion in new investments.

The measure was opposed by the Clean Affordable Renewable Energy for Michigan Coalition, or C.A.R.E., as well as Governor Rick Snyder, the Michigan Chamber of Commerce, and Citizens Protecting Michigan's Constitution. The opponents pointed out the inherent difficulty in meeting the high standards set by the amendment, which would be the only such mandate in a state constitution. In particular, Citizens Protecting Michigan's constitution launched a Tea Party–inspired "no change to the constitution" battle against all six of the measures that appeared on the 2012 ballot in November, five of which were constitutional amendments. The business community united around a "vote no on everything" campaign strategy.

In a September 2012 video press release, Governor Snyder, voicing a common argument for the opposition, said,

> Current law sets a goal of generating 10 percent of electricity from renewable sources such as wind, solar, hydro and biomass by 2015. This is a standard that's already difficult to meet. Proposal Three would set the bar even higher—and we would be the only state to have such a mandate in our constitution.[55]

Michigan Energy, Michigan Jobs, the main committee formed to promote the amendment, spent nearly $14 million, with other partners ponying up an additional $8 million in support. The committees formed in opposition spent close to $32 million, with C.A.R.E. leading the way, spending $24.5 million to defeat the amendment.[56] Facing enormous spending against it and vocal opposition by the governor, as well as a broader anti–ballot measure mood, Michigan's Renewable Energy Amendment ultimately failed, winning only 38 percent of the vote.

Conclusion

Since 1908, environmental groups have leveraged direct democracy mechanisms to protect air, water, land, and wildlife; to regulate oil and gas emissions; and to promote renewable energy. In this chapter, we have shown that despite significant variance in passage rates across categories of environmental protection, the mechanisms of direct democracy do offer proenvironment activists an additional venue to achieve their goals. We have also established that renewable energy measures, which deal most directly with efforts to stave off climate change, are not significantly less likely to pass than most other environmental measures, all else equal. Most important for climate change activists, we have shown that institutional factors have a significant impact on passage rates for all types of environmental measures. Specifically, referred measures are significantly more likely to pass than citizen initiatives.

These findings suggest that previous research may be too hasty in raising concerns about the capacity of environmental measures to overcome the barriers created by opposing coalitions of business interests with deep pockets. To be sure, not all proenvironment ballot measures succeed at the polls, even those that are well financed and professionally run, as was the case in 2012 in Michigan with Proposal 3. But for more than a century, proenvironment forces have had numerous victories at the polls, especially legislative referendums and measures on ballots in elections in which ballot measures had high levels of overall support. Many of these ballot measures—particularly measures dealing with air and water quality—achieve success by having bipartisan support and by staying below the radar. This is welcome news for advocates thinking about waging a proenvironment ballot initiative but who lack the financial resources of their corporate opponents—at least those working in states whose legislatures may be willing to place proenvironment measures on the ballot. But it also means that environmental activists will have to work harder to lobby lawmakers to place legislative referrals on the ballot.

Considering the rather tepid response of Congress to the issue of climate change and other environmental issues, placing proenvironment initiatives and referendums before voters at the ballot box, while by no means an electoral certainty, offers environmental activists another avenue to build on their past successes. American proenvironment activists, including those especially concerned with issues of climate change, could take a cue from citizens in many European countries, who have successfully leveraged direct democracy institutions at the local level to support climate protection. In the United States, there is no reason not to expect in the future more renewable energy measures on state and local ballots, including those that clearly and unselfconsciously invoke the language of climate change to appeal to voters. As Giddens argues, because markets alone cannot provide a solution, state intervention is necessary to confront the externalities of climate change, yet, a systematic politics of climate change does not exist.[57] While by no means perfect, the mechanisms of direct democracy do offer some hope to those who are interested in confronting the inevitable environmental consequences stemming from climate change.

Suggested Readings

Bowler, Shaun, and Todd Donovan. *Demanding Choices: Opinion, Voting, and Direct Democracy.* Ann Arbor: University of Michigan Press, 1998.

Chong, Dennis, and Yael Wolinsky-Nahmias. "Managing Voter Ambivalence in Growth and Conservation Campaigns." In *Ambivalence, Politics, and Public Policy,* edited by Stephen C. Craig and Michael D. Martinez, 103–125. New York: Palgrave Macmillan, 2005.

Guber, Deborah Lynn. "Environmental Voting in the American States: A Tale of Two Initiatives." *State and Local Government Review* 33 (2001): 120–132.

Magleby, David B. *Direct Legislation: Voting on Ballot Propositions in the United States.* Baltimore: Johns Hopkins University Press, 1984.

References

Bord, Richard J., Ann Fisher, and Robert E. O'Connor. "Public Perceptions of Global Warming: United States and International Perspectives." *Climate Research* 11 (1998): 73–84.

Bornstein, Nicholas, and Bruno Lanz. "Voting on the Environment: Price or Ideology? Evidence From Swiss Referendums." *Ecological Economics* 67 (2008): 430–440.

Bowler, Shaun, and Todd Donovan. *Demanding Choices: Opinion, Voting, and Direct Democracy.* Ann Arbor: University of Michigan Press, 1998.

Bowler, Shaun, Todd Donovan, and Trudi Happ. "Ballot Propositions and Information Costs: Direct Democracy and the Fatigued Voter." *Western Political Quarterly* 45, no. 2 (1992): 559–568.

Branton, Regina P. "Examining Individual-Level Voting Behavior on State Ballot Propositions." *Political Research Quarterly* 56 (2003): 367–377.

Chong, Dennis, and Yael Wolinsky-Nahmias. "Managing Voter Ambivalence in Growth and Conservation Campaigns." In *Ambivalence, Politics, and Public Policy,* edited by Stephen C. Craig and Michael D. Martinez, 103–126. New York: Palgrave Macmillan, 2005.

Daniels, David P., and Jon A. Krosnick, Michael P. Tichy, and Trevor Tompson. "Public Opinion on Environmental Policy in the United States." In *The Oxford Handbook of U.S. Environmental Policy,* edited by Sheldon Kamieniecki and Michael Kraft, 461–486. New York: Oxford University Press, 2012.

Davis, Frank L., and Albert H. Wurth. "Voting Preferences and the Environment in the American Electorate: The Discussion Extended." *Society and Natural Resources: An International Journal* 16 (2003): 729–740.

De Figueiredo, John M., Chang Ho Ji, and Thad Kousser. "Financing Direct Democracy: Revisiting the Research on Campaign Spending and Citizen Initiatives." *Journal of Law, Economics, and Organization* 27 (2011): 485–514.

Dealbook. "Pickens-Backed Bill Is Shot Down in California." *New York Times,* November 5, 2008.

Dunlap, Riley E. "Trends in Public Opinion Toward Environmental Issues: 1965–1990." *Society and Natural Resources: An International Journal* 4 (1991): 285–312.

Dunlap, Riley E., and Richard P. Gale. "Party Membership and Environmental Politics: A Legislative Roll-Call Analysis." *Social Science Quarterly* 55 (1974): 670–690.

Dunlap, Riley E., and Rik Scarce. "The Polls—Poll Trends: Environmental Problems and Protection." *Public Opinion Quarterly* 55 (1991): 651–672.

Dunlap, Riley E., Chenyang Xiao, and Aaron M. McCright. "Politics and Environment in America: Partisan and Ideological Cleavages in Public Support for Environmentalism." *Environmental Politics* 10 (2001): 23–48.

Elliot, Euel, James L. Regens, and Barry J. Seldon. "Exploring Variation in Public Support for Environmental Protection." *Social Science Quarterly* 76 (1995): 41–52.

Garrett, Elizabeth, and Daniel A. Smith. "Veiled Political Actors and Campaign Disclosure Laws in Direct Democracy." *Election Law Journal* 4 (2005): 295–328.

Gerlak, Andrea K., and Susan M. Natali. *Taking the Initiative II.* Washington, DC: Americans for the Environment, 1993.

Giddens, Anthony. *The Politics of Climate Change,* 2nd ed. Cambridge, UK: Polity, 1993.

Goebel, Thomas. *A Government by the People: Direct Democracy in America, 1890–1940.* Chapel Hill: University of North Carolina Press, 2002.

Governor Rick Snyder. "Governor Snyder Speaks Out: Michigan's 2012 Ballot Initiatives." http://michigan.gov/snyder/0,4668,7-277-60279-286346—,00.html.

Guber, Deborah Lynn. "Environmental Voting in the American States: A Tale of Two Initiatives." *State and Local Government Review* 33 (2001): 120–132.

Guber, Deborah Lynn. "Voting Preferences and the Environment in the American Electorate." *Society and Natural Resources* 14 (2001): 455–469.

Lake, Laura M. "The Environmental Mandate: Activists and the Electorate." *Political Science Quarterly* 98 (1983): 215–233.

Lupia, Arthur. "Shortcuts Versus Encyclopedias: Information and Voting Behavior in California Insurance Reform Elections." *American Political Science Review* 88 (1994): 63–76.

Lutrin, Carl E., and Allen K. Settle. "The Public and Ecology: The Role of Initiatives in California's Environmental Politics." *Western Political Quarterly* 28 (1975): 352–371.

Magleby, David B. *Direct Legislation: Voting on Ballot Propositions in the United States.* Baltimore: Johns Hopkins University Press, 1984.

Mertig, Angela G., and Riley E. Dunlap. "Public Approval of Environmental Protection and Other New Social Movement Goals in Western Europe and the United States." *International Journal of Public Opinion Research* 7 (1995): 145–156.

National Institute on Money and State Politics. http://www.followthemoney.org/.

Nicholson, Stephen P. "The Political Environment and Ballot Proposition Awareness." *American Journal of Political Science* 47 (2003): 403–410.

Nicholson, Stephen P. *Voting the Agenda: Candidates, Elections, and Ballot Propositions.* Princeton, NJ: Princeton University Press, 2005.

Petrocik, John R. "Issue Ownership in Presidential Elections, With a 1980 Case Study." *American Journal of Political Science* 40 (1996): 825–850.

Pew Research Center for the People and the Press. "Public Priorities: Deficit Rising, Terrorism Slipping." January 23, 2012. http://www.people-press.org/2012/01/23/public-priorities-deficit-rising-terrorism-slipping/.

Schuldt, Jonathon A., Sara H. Konrath, and Norbert Schwarz. "'Global Warming' or 'Climate Change'? Whether the Planet Is Warming Depends on Question Wording." *Public Opinion Quarterly* 75 (2011): 115–124.

Smith, Daniel A. "Campaign Financing of Ballot Initiatives in the American States." In *Dangerous Democracy? The Battle Over Ballot Initiatives in America,* edited by Larry Sabato, Bruce Larson, and Howard Ernst, 71–90. Lanham, MD: Rowman and Littlefield, 2011.

Smith, Daniel A. "Direct Democracy and Candidate Elections," in *The Electoral Challenge: Theory Meets Practice,* 2nd ed., edited by Stephen C. Craig and David Hill, 194–208. Washington, DC: CQ Press, 2010.

Smith, Daniel A. "Direct Democracy: Regulating the 'Will of the People.'" In *Law and Election Politics: The Rules of the Game,* 2nd ed., edited by Matthew J. Streb, 171–190. New York: Routledge, 2012.

Smith, Daniel A., and Caroline. J. Tolbert. *Educated by Initiative: The Effects of Direct Democracy on Citizens and Political Organizations in the United States.* Ann Arbor: University of Michigan Press, 2004.

Smith, Daniel A., and Caroline J. Tolbert. "The Initiative to Party: Partisanship and Ballot Initiatives in California." *Party Politics* 7 (2001): 739–757.

Stratmann, Thomas. "Campaign Spending and Ballot Measures." In *Financing Referendum Campaigns,* edited by Karin Gilland Lutz and Simon Hug, 9–22. New York: Palgrave Macmillan, 2010.

Stratmann, Thomas. "Is Spending More Potent For or Against a Proposition? Evidence From Ballot Measures." *American Journal of Political Science* 50 (2006): 788–801.

Villar, Ana, and Jon A. Krosnick. "Global Warming vs. Climate Change, Taxes vs. Prices: Does Word Choice Matter?" *Climatic Change* 105 (2011): 1–12.

Yeager, David Scott, Samuel B. Larson, Jon A. Krosnick, and Trevor Tompson. "Measuring Americans' Issue Priorities: A New Version of the Most Important Problem Question Reveals More Concern About Global Warming and the Environment." *Public Opinion Quarterly* 75 (2011): 125–138.

Young, Samantha. "California Rejects Clean Power Initiative." November 5, 2008. Associated Press.

Zaller, John R. *The Nature and Origins of Mass Opinion.* New York: Cambridge University Press, 1992.

Notes

1. Thomas Goebel, *A Government by the People: Direct Democracy in America, 1890–1940* (Chapel Hill: University of North Carolina Press, 2002).

2. David B. Magleby, *Direct Legislation: Voting on Ballot Propositions in the United States* (Baltimore: Johns Hopkins University Press, 1984).

3. John R. Zaller, *The Nature and Origins of Mass Opinion* (New York: Cambridge University Press, 1992).

4. Riley E. Dunlap and Rik Scarce, "The Polls—Poll Trends: Environmental Problems and Protection," *Public Opinion Quarterly* 55 (1991): 651–672.

5. Pew Research Center for the People and the Press, "Public Priorities: Deficit Rising, Terrorism Slipping," January 23, 2012, http://www.people-press.org/2012/01/23/public-priorities-deficit-rising-terrorism-slipping/.

6. Jonathon A. Schuldt, Sara H. Konrath, and Norbert Schwarz, "'Global Warming' or 'Climate Change'"? Whether the Planet Is Warming Depends on Question Wording," *Public Opinion Quarterly* 75 (2011): 115–124; Euel Elliot, James L. Regens, and Barry J. Seldon, "Exploring Variation in Public Support for Environmental Protection," *Social Science Quarterly* 76 (1995): 41–52.

7. Riley E. Dunlap and Richard P. Gale, "Party Membership and Environmental Politics: A Legislative Roll-Call Analysis," *Social Science Quarterly* 55 (1974): 670–690.

8. Riley E. Dunlap, Chenyang Xiao, and Aaron M. McCright, "Politics and Environment in America: Partisan and Ideological Cleavages in Public Support for Environmentalism," *Environmental Politics* 10 (2001): 30.

9. Deborah Lynn Guber, "Voting Preferences and the Environment in the American Electorate," *Society and Natural Resources* 14 (2001): 455–469.

10. Frank L. Davis and Albert H. Wurth, "Voting Preferences and the Environment in the American Electorate: The Discussion Extended," *Society and Natural Resources: An International Journal* 16 (2003): 729–740.

11. Angela G. Mertig and Riley E. Dunlap, "Public Approval of Environmental Protection and Other New Social Movement Goals in Western Europe and the United States," *International Journal of Public Opinion Research* 7 (1995): 145–156.

12. Dunlap and Scarce, "The Polls."

13. Ibid.

14. Richard J. Bord, Ann Fisher, and Robert E. O'Connor, "Public Perceptions of Global Warming: United States and International Perspectives," *Climate Research* 11 (1998): 73–84.

15. Magleby, *Direct Legislation.*

16. Laura M. Lake, "The Environmental Mandate: Activists and the Electorate," *Political Science Quarterly* 98 (1983): 215–233; Deborah Lynn Guber, "Environmental Voting in the American States: A Tale of Two Initiatives," *State and Local Government Review* 33 (2001): 120–132.

17. Carl E. Lutrin and Allen K. Settle, "The Public and Ecology: The Role of Initiatives in California's Environmental Politics," *Western Political Quarterly* 28 (1975): 352–371; Magleby, *Direct Legislation.*

18. Arthur Lupia, "Shortcuts Versus Encyclopedias: Information and Voting Behavior in California Insurance Reform Elections," *American Political Science Review* 88 (1994): 63–76.

19. Daniel A. Smith and Caroline J. Tolbert, "The Initiative to Party: Partisanship and Ballot Initiatives in California," *Party Politics* 7 (2001): 739–757; Smith and Tolbert,

Educated by Initiative: The Effects of Direct Democracy on Citizens and Political Organizations in the United States (Ann Arbor, University of Michigan Press, 2004); Regina P. Branton, "Examining Individual-Level Voting Behavior on State Ballot Propositions," *Political Research Quarterly* 56 (2003): 367–377.

20. David P. Daniels, Jon A. Krosnick, Michael P. Tichy, and Trevor Tompson, "Public Opinion on Environmental Policy in the United States," in *The Oxford Handbook of U.S. Environmental Policy*, ed. Sheldon Kamieniecki and Michael Kraft (New York: Oxford University Press, 2012), 461–486; Ana Villar and Jon A. Krosnick, "Global Warming vs. Climate Change, Taxes vs. Prices: Does Word Choice Matter?" *Climatic Change* 105 (2011): 1–12; David Scott Yeager, Samuel B. Larson, Jon A. Krosnick, and Trevor Tompson, "Measuring Americans' Issue Priorities: A New Version of the Most Important Problem Question Reveals More Concern About Global Warming and the Environment," *Public Opinion Quarterly* 75 (2011): 125–138; Dunlap, Xiao, and McCright, "Politics and Environment in America."

21. Daniel A. Smith, "Direct Democracy and Candidate Elections," in *The Electoral Challenge: Theory Meets Practice*, 2nd ed., ed. Stephen C. Craig and David B. Hill (Washington, DC, CQ Press, 2010), 194–208.

22. Magleby, *Direct Legislation.*

23. Ibid.; Dennis Chong and Yael Wolinsky-Nahmias, "Managing Voter Ambivalence in Growth and Conservation Campaigns," in *Ambivalence, Politics, and Public Policy*, ed. Stephen C. Craig and Michael D. Martinez (New York: Palgrave Macmillan, 2005), 103–125.

24. Lake, "The Environmental Mandate."

25. Andrea K. Gerlak and Susan M. Natali, *Taking the Initiative II* (Washington, DC: Americans for the Environment, 1993).

26. Magleby, *Direct Legislation*; Daniel A. Smith, "Campaign Financing of Ballot Initiatives in the American States," in *Dangerous Democracy? The Battle Over Ballot Initiatives in America*, ed. Larry Sabato, Bruce Larson, and Howard Ernst (Lanham, MD: Rowman and Littlefield, 2001), 71–90.

27. Daniel A. Smith, "Direct Democracy: Regulating the 'Will of the People,'" in *Law and Election Politics: The Rules of the Game*, 2nd ed., ed. Matthew J. Streb (New York: Routledge, 2012), 171–190.

28. Elizabeth Garrett and Daniel A. Smith, "Veiled Political Actors and Campaign Disclosure Laws in Direct Democracy," *Election Law Journal* 4 (2005): 295–328.

29. John M. De Figueiredo, Chang Ho Ji, and Thad Kousser, "Financing Direct Democracy: Revisiting the Research on Campaign Spending and Citizen Initiatives," *Journal of Law, Economics, and Organization* 27 (2011): 485–514.

30. Thomas Stratmann, "Is Spending More Potent For or Against a Proposition? Evidence From Ballot Measures," *American Journal of Political Science* 50 (2006): 788–801.

31. Lutrin and Settle, "The Public and Ecology."

32. Guber, "Environmental Voting."

33. Chong and Wolinsky-Nahmias, "Managing Voter Ambivalence."

34. Ibid., 103–104.

35. Gerlak and Natali, *Taking the Initiative II.*

36. Chong and Wolinsky-Nahmias, "Managing Voter Ambivalence," 121.

37. Nicholas Bornstein, and Bruno Lanz, "Voting on the Environment: Price or Ideology? Evidence From Swiss Referendums." *Ecological Economics* 67 (2008): 430–440.

38. Lake, "The Environmental Mandate,"

39. Magleby, *Direct Legislation.*

40. Shaun Bowler and Todd Donovan, *Demanding Choices: Opinion, Voting, and Direct Democracy* (Ann Arbor: University of Michigan Press, 1998); Stephen P. Nicholson, "The Political Environment and Ballot Proposition Awareness," *American Journal of Political Science* 47 (2003): 403–410.

41. Shaun Bowler, Todd Donovan, and Trudi Happ, "Ballot Propositions and Information Costs: Direct Democracy and the Fatigued Voter," *Western Political Quarterly* 45, no. 2 (1992): 559–568.

42. Lake, "The Environmental Mandate."

43. Bowler and Donovan, *Demanding Choices;* Stephen P. Nicholson, *Voting the Agenda: Candidates, Elections, and Ballot Propositions* (Princeton, NJ: Princeton University Press, 2005).

44. Bowler and Donovan, *Demanding Choices.*

45. Ibid.

46. Chong and Wolinsky-Nahmias, "Managing Voter Ambivalence."

47. Schuldt, Konrath, and Schwarz, "'Global Warming' or 'Climate Change'?"

48. Ibid.

49. National Institute on Money in State Politics, "Citizens for Sensible Energy Choices," http://www.followthemoney.org/database/StateGlance/committee.phtml?c=1152; National Institute on Money in State Politics, "Intermountain Rural Electric Association," http://www.followthemoney.org/database/StateGlance/committee.phtml?c=1161.

50. Chong and Wolinsky-Nahmias, "Managing Voter Ambivalence"; Gerlak and Natali, *Taking the Initiative II;* Guber, "Environmental Voting."

51. National Institute on Money in State Politics, "Citizens for Sensible Energy Choices," "Intermountain Rural Electric Association."

52. Samantha Young, "California Rejects Clean Power Initiative," Associated Press, November 5, 2008.

53. "No on Proposition 10," http://digital.library.ucla.edu/websites/2008_993_114/www.noonproposition10.org/.

54. Dealbook, "Pickens-Backed Bill Is Shot Down in California," *New York Times,* November 5, 2008.

55. Governor Rick Snyder, "Governor Snyder Speaks Out: Michigan's 2012 Ballot Initiatives," November 5, 2008, http://michigan.gov/snyder/0,4668,7–277–60279–286346—,00.html.

56. National Institute on Money in State Politics, "Proposal 12-3: Electricity From Renewable Energy Sources," http://www.followthemoney.org/database/StateGlance/ballot.phtml?m=1026&utm_campaign=following-the-money-summer-2012&utm_medium=email&utm_source=nimsp-contacts.

57. Anthony Giddens, *The Politics of Climate Change,* 2nd ed. (Cambridge, UK: Polity Press, 1993).

CHAPTER 8

Consumer Political Action on Climate Change

Lauren Copeland and Eric R. A. N. Smith

Introduction

In the spring of 2003, Carrotmob founder Brent Schulkin got an idea: What if, instead of boycotting companies for poor environmental practices, citizens worked with companies to mitigate the effects of climate change?[1] As Schulkin saw it, the government was not doing much to address problems associated with climate change, so why not turn to other powerful actors like businesses and corporations? He knew companies would not participate voluntarily, however, if any new regulations or efforts to stem air and water pollution cut into their bottom line. Therefore, he needed to incentivize companies to improve their environmental practices.

Over the next few years, Schulkin realized that he could incentivize companies by promising to deliver a critical mass (or flashmob) of consumers to the business that agreed to commit the highest percentage of profits it received from these consumers to a proenvironment action. On March 29, 2008, Schulkin's idea came to fruition. That day, hundreds of green-minded consumers spent more than $9,200 at K&D Market, which in turn spent 22 percent of the day's profits on energy efficiency upgrades. Since then, more than 300 Carrotmob campaigns have occurred worldwide. As a result of these campaigns, companies have improved their environmental practices by retrofitting their infrastructures to be more energy efficient, cutting electricity usage, sourcing from local suppliers, and installing solar panels to generate electricity, among other actions.

When people participate in Carrotmobs, they are acting as political consumers, or people who effect social and political change through their purchasing behavior. In this chapter, we introduce political consumerism as an innovative political tool people use to address environmental problems such as climate change. In what follows, we provide an overview of political consumerism. Next, we discuss differences between political consumers and nonpolitical consumers, including which types of people are more likely to engage in political consumerism than others. In the third section, we explore the roles organizations play in boycotts and buycotts. Then, we turn to the role of political consumerism in climate change politics. Finally, we evaluate the extent to which political consumerism is an effective tool for achieving social and political change. Overall, we conclude that political consumerism is an effective way to promote more environmentally sustainable practices in the marketplace, especially until sufficient regulatory and enforcement mechanisms are in place.

Political Consumerism: An Overview

Political consumerism refers to decisions to avoid or seek specific products for political, ethical, or environmental reasons with the goal of changing objectionable market or institutional practices. By engaging in boycotts and buycotts—which collectively compose political consumption—people can push their social and political agendas. Through boycotts, people can punish companies for undesirable behavior. At the same time, people can buycott, or purchase specific products or brands, to reward companies for favorable behavior. For example, people may boycott ExxonMobil because the company funds research to disprove anthropogenic, or human-caused, climate change. Similarly, people may buycott hybrid cars to make political statements.

Boycotting and buycotting are not new political tools. In fact, they are at least as old as the boycotts of British goods in the US prerevolutionary period. However, the frequency with which people engage in boycotting and buycotting has increased over time. According to the World Values Survey, the percentage of Americans and Europeans who engage in boycotts increased from 5 to 15 percent between 1974 and 1999. The most recent study of political consumerism in the United States finds that 47 percent of people engage in political consumerism; of these, 24 percent engage in boycotts, 18 percent in buycotts, and 58 percent in both.[2]

Most of the literature attributes the expansion of political consumerism to the rise of postmaterialist values in industrialized nations in the decades following World War II. According to one prominent scholar on value change and political behavior, cohorts who were socialized during the postwar era were more likely to emphasize postmaterialist or quality-of-life concerns, such as the environment and civil rights.[3] Moreover, they preferred to engage in direct, or elite-challenging, forms of political action like strikes, marches, and boycotts, because they valued autonomy and self-expression.

People also turn to direct forms of political action like political consumerism when governments cannot address political problems effectively. Andersen and Tobiasen explain, "Within the framework of globalization, political consumerism takes on a particular significance, sometimes providing the only opportunity to influence outcomes as transnational companies are outside the regulatory powers of the national governments."[4] Unlike economic problems, which can be addressed by local, state, and national governments, environmental problems are difficult to solve, because those who bear the environmental burden of their effects are often in a different region from those who cause them. Thus, political consumerism provides a vehicle through which citizens can effect change until global structures with sufficient regulatory and enforcement power are in place.

The inability of governments to achieve broad-based collective action to address problems associated with environmental degradation helps explain why people turn to direct forms of political action to address social and political grievances. We can see evidence of people turning to political consumerism in a nationally representative survey of US adults that was conducted by one of the authors.[5] These data are displayed in Figure 8.1.

As Figure 8.1 shows, 14 percent of boycotters—and 17 percent of buycotters—engage in political consumerism for environmental reasons. In addition, 39 percent

FIGURE 8.1   Percentage of US Adults Who Boycotted or Buycotted Any Product or Corporation in 2011

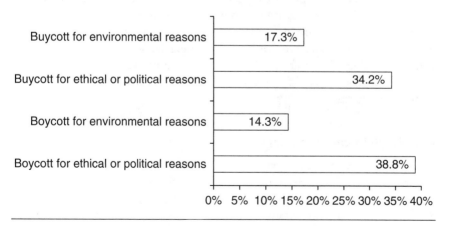

Source: Original survey data collected in the United States between December 7 and 21, 2011.

of people who engage in boycotts—and 34 percent of people who engage in buycotts—do so for ethical or political reasons.

Other studies show that people who score higher on measures of environmental concern are more likely to engage in political consumerism. Researchers have found that proenvironment beliefs are strong predictors of political consumerism, including buying organic fruits and vegetables, products made from recycled materials, and household cleaners that are environmentally friendly. People who scored higher on measures of environmental concern were also more likely to boycott BP after the 2010 *Deepwater Horizon* oil spill, during which 4.9 million barrels of oil seeped into the Gulf of Mexico. Still other studies show that political discussion and environmental concern are significantly and positively related to political consumerism. Finally, political consumers are significantly more likely to belong to environmental organizations than are nonpolitical consumers.

People also engage in political consumerism because they believe it is their responsibility to be conscious consumers. Figure 8.2 compares the percentages of political consumers and nonpolitical consumers who agree that boycotting and buycotting are meaningful activities.

While 61 percent of political consumers believe "it is the personal responsibility of each citizen to purchase products and services from ethical companies," only 23 percent of nonpolitical consumers agree with this statement. Similarly, while 63 percent of political consumers believe "boycotting and buycotting are forms of political expression," only 43 percent of nonpolitical consumers feel the same way. Finally, political consumers are much more likely to believe that political consumerism matters. Whereas 76 percent of political consumers believe "citizens can influence society by purchasing goods and services from ethical companies," only 36 percent of nonpolitical consumers agree with this statement. In short, people who engage in boycotts and buycotts are much more likely than are nonpolitical consumers to believe that their purchasing decisions should be consistent with their values. Political consumers are also more likely than nonpolitical consumers to believe that they are engaging in political participation and that their participation matters.

Who Are Political Consumers?

Thus far, we have addressed why people engage in political consumerism, but we have not addressed which types of people are more likely to engage in political consumerism than others. We can learn which types of people are more likely to engage in political consumerism than others by drawing once again on our national survey data. Table 8.1 compares people who do not engage in political

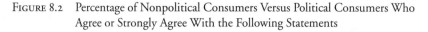

FIGURE 8.2   Percentage of Nonpolitical Consumers Versus Political Consumers Who
Agree or Strongly Agree With the Following Statements

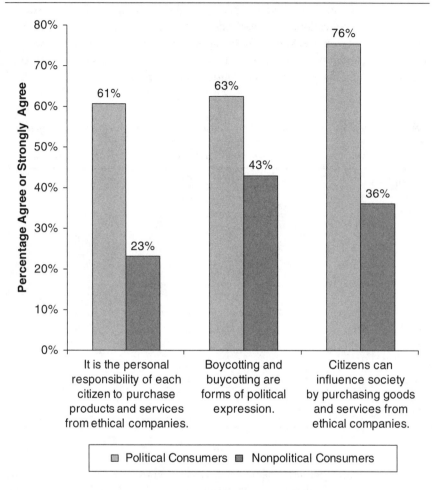

Source: Original survey data collected in the United States between December 7 and 21, 2011.

consumerism to people who engage in boycotting or buycotting. Each
respondent was asked to report his or her age in years, highest level of education
completed, which ranged from 1 ("no high school degree") to 6 ("postgraduate
degree"), and annual household income, which ranged from 1 ("less than
$10,000") to 14 ("$150,000 or more").* The entry in each cell represents the

mean score of all individuals surveyed, except for age, where we used the mean of all ages reported.

There are significant differences between political and nonpolitical consumers with respect to education, income, gender, external efficacy, strength of ideology, and political interest. On average, political consumers are better educated, wealthier, and more interested in politics than are nonpolitical consumers. Political consumers are more likely to be male.[6] Political consumers also have higher levels of

TABLE 8.1   Descriptive Statistics of Nonpolitical Consumers and Political Consumers

Consumer characteristic	Nonpolitical consumers	Political consumers
Age in years	47	46
Education	2.8	3.6
Income	6.7	8.0
Male	0.49	0.54
Political interest	1.9	2.50
Internal efficacy	1.4	1.4
External efficacy	1.9	2.1
Partisan strength	2.0	1.9
Ideological strength	0.8	1.1

---

* Gender was coded 1 for male. Political interest was captured by a single question asking respondents how often they paid attention to the news and public affairs, and it was measured on a four-point scale ranging from 0 ("hardly at all") to 3 ("most of the time"). Internal efficacy was measured by asking respondents how much they agreed with the following statement: "Public officials care about what people like me think." The scale ranged from 0 ("strongly disagree") to 4 ("strongly agree"). External political efficacy was measured by asking respondents to indicate how much they felt they could affect what the government does. The scale ranged from 0 ("not at all") to 4 ("a great deal"). Partisan strength is a folded four-category measure of the standard seven-point party identification scale, ranging from pure independents ("0") to strong party identifiers ("3"). Similarly, ideological strength is a folded three-category measure of the standard five-point ideology scale, ranging from 0 ("moderate") to 2 ("very liberal or very conservative").

external efficacy; they are more likely than others to believe they can "affect what the government does." Finally, political consumers are more likely to identify themselves as very liberal or very conservative, but they are less likely to identify with a political party.

Political consumers are also different from nonpolitical consumers with respect to their sociopolitical attitudes and beliefs, which are shown in Table 8.2. Again, cell entries are mean scores. Higher scores indicate higher levels of trust, postmaterialist values, and proenvironment beliefs. *

Compared to nonpolitical consumers, political consumers are more likely to believe other people can be trusted, but they have lower levels of political trust; they are less likely to believe "you can trust the government in Washington to do what is right," and they are more likely to believe that "most of the people running the government are corrupt." Having lower levels of political trust is consistent with the idea that people turn to unconventional and direct forms of political participation when they feel that government cannot address their concerns.

Newman and Bartels explain why direct forms of political action, such as political consumerism, appeal to people with low levels of political trust: "Distrust toward government increases the perceived value and efficacy of extraelectoral and noninstitutionalized routes of citizen influence relative to those organized and mediated by government."[7] Consequently, political consumers prefer direct forms of political action, as well as participation in groups that advocate for civil rights,

TABLE 8.2   Sociopolitical Attitudes of Nonpolitical Consumers and Political Consumers

Attitude	Nonpolitical consumers	Political consumers
Social trust	3.46	4.02
Political trust	6.47	5.62
Postmaterialism	1.00	1.37
Environmentalism	11.54	12.72

* Social trust was measured by asking people the following question: "Generally speaking, would you say that most people can be trusted or that you need to be very careful in dealing with people?" The scale ranged from 1 ("most people can be trusted") to 10 ("need to be very careful") and was reverse scored so that higher values indicate more trust. Political trust was measured using two items as an additive scale ranging from 0 to 18, with higher scores indicating more trust. The postmaterialism scale ranged from 0 ("materialist") to 2 ("postmaterialist"). Finally, environmentalism was measured using a 20-point scale, with higher scores indicating more proenvironment beliefs.

third-world cooperation, and environmentalist, pacifist, and consumer causes. Finally, compared to nonpolitical consumers, political consumers score significantly higher on measures of postmaterialism and environmentalism, just as we would expect.

One might wonder whether people who engage in boycotting and buycotting also engage in other types of civic and political participation. On the one hand, critics of political consumerism argue that boycotting and buycotting individualize political action and relieve the government and other institutions of responsibility for implementing public policy to regulate the marketplace. As Maniates puts it,

> When confronted by environmental ills—ills many confess to caring deeply about—Americans seem capable of understanding themselves almost solely as consumers who must buy "environmentally sound" products (and then recycle them), rather than as citizens who might come together and develop political clout sufficient to alter institutional arrangements that drive a pervasive consumerism.[8]

In other words, if people understand environmental degradation as a consequence of individual actions, then they are more likely to believe that they should take responsibility for reversing environmental ills—by, for example, boycotting and buycotting—instead of calling on political institutions to implement change.

On the other hand, political consumerism may be part of a larger participation repertoire, so that people who engage in political consumerism also engage in other forms of political participation directed at the state. We explore this issue in Table 8.3. We assigned scores to respondents based on their levels of participation as indicated in our survey. The scores ranged from 0 to 4. Here, higher scores indicate higher levels of participation.

Table 8.3 includes four main types of participation: electoral participation, individualistic participation, civic participation, and online participation. *Electoral*

TABLE 8.3   Civic and Political Participation by Nonpolitical Consumers and Political Consumers

Type of participation	Nonpolitical consumers	Political consumers
Electoral participation	0.21	0.64
Individualistic participation	0.63	1.65
Civic participation	0.48	1.08
Online participation	0.09	0.31

participation includes working or volunteering for a political group or cause and attending a political meeting or rally. *Individualistic participation* refers to signing a petition, contacting a political or public official, and donating money. *Civic participation* includes working or volunteering for a nonpolitical group or cause and working with others in the community to solve a problem. Finally, *online participation* includes following or friending a candidate or political group on a social networking site, as well as following or friending a civic organization or nonprofit group on a social networking site. The results show that political consumers are significantly more likely than nonpolitical consumers to engage in all types of civic and political participation. Political consumers are especially more likely to engage in individualistic and online forms of participation. This shows that boycotting and buycotting complement a larger participatory repertoire. People do not engage in political consumerism at the expense of other civic and political activities.

These findings are consistent with another study, in which Strømsnes finds a strong, positive relationship between political consumerism and 13 political acts. She also finds that people who engage in political consumerism are generally more likely to participate in politics.[9] Other studies also find that political consumers are more likely than non political consumers to contact public officials, participate in government hearings, and write letters to the editor. Collectively, these results suggest that people who engage in boycotts and buycotts understand that large-scale change requires government involvement.

Organizing Political Consumers

Buycotts and boycotts happen for a variety of reasons. In many cases, political groups organize and lead them. In some cases, governments try to influence people's purchasing behavior (e.g., "Buy American"). In other cases, people make individual decisions about what to buy or avoid for their own ethical or political reasons without guidance from any organized campaign. Still, in order to understand an important segment of political consumerism, one must look at those who organize boycotting or buycotting campaigns and consider their methods.

The research on boycotts distinguishes between marketplace-oriented and media-oriented boycotts. The former seek to achieve their goals by harming a firm's sales; the latter try to achieve their goals by harming both a firm's sales and its public image. Some boycotts are more successful than others, for several reasons.[10] For boycotts aimed at harming a company's sales, both the products that organizers want consumers to stop buying and the brand names should be easily identifiable to consumers. Many organizations have launched campaigns to protect rainforests because rainforests provide an important carbon sink that keeps carbon dioxide (CO_2) out of the atmosphere. However, some campaigns have been more successful than others. For example, when the Rainforest Action Network (RAN) launched its effort to stop logging in old growth forests, it asked consumers not to buy wood at Home Depot stores. The campaign succeeded, in part, because it was a simple message that was easy to remember.

Organizers should also run their campaigns when there are no other competing campaigns aimed at similar products. When different activist groups target similar companies at the same time, they confuse the public, making it difficult for any one group to achieve its objectives. As of this writing, for instance, all major oil companies that do retail business in the United States, except ConocoPhillips, are targets of boycotts. The Green Party is running a campaign to boycott ExxonMobil, Texaco, and Chevron. The Defenders of Wildlife, Greenpeace, the National Resources Defense Council, the Sierra Club, the US Public Interest Research Group (US PIRG), and the Union of Concerned Scientists are running the Exxpose Exxon campaign against ExxonMobil. In addition, Public Citizen is using a Facebook campaign to boycott BP (which formerly labeled itself British Petroleum), and several independent groups are trying to boycott Shell Oil. In large part, these efforts are failing to change any of the oil companies' behavior. A single, cohesive campaign to pressure one oil company to modify its behavior would be much more effective.

The oil company boycotts bring up another point: Boycotts are more likely to be successful when there are acceptable substitutes for the targets of boycotts. When organizers target every major oil company, consumers have no alternative company from which they can purchase gasoline for their automobiles. In contrast, RAN's boycott of Home Depot worked, because the organization did not ask people to avoid buying wood or shopping at home improvement centers, only to boycott Home Depot. After the company agreed to stop selling old growth lumber, RAN switched its campaign to another target, Lowe's, which also gave in to RAN's demands.

Finally, consumer violations of boycotts should be publicly visible. Much like crossing a picket line in a labor dispute, publicly displaying a violation of a boycott—such as driving a gas-guzzling Hummer—can draw negative attention, and some people may feel that it will harm their personal reputations. That is why a boycott of a product that must be displayed in public, such as a car, is likely to be more successful than a boycott of a product that can be purchased and used in private.

Market-oriented and media-oriented boycotts have several characteristics in common. Through media-oriented boycotts, organizers threaten a company's reputation to win concessions. Boycotts that focus on firms' public images may succeed even when the boycott has little impact on sales, because businesses regard their public images as important investments, and they want to protect them. In fact, many board members live in fear "of getting their corporate reputations blown away" and may yield to a boycott in order to avoid doing so.[11] This is why media-oriented boycotts can be just as effective as market-oriented boycotts. For example, when RAN threatened to boycott Disney for using paper products made from wood that came from clear-cut Indonesian rainforests, the entertainment giant negotiated a comprehensive paper policy that applied to all of its operations in the United States and internationally, and even to the licensees of Disney characters as well.[12] Disney's decision made good business sense. After all, they had good reasons to fear a group that could start a boycott with slogans such as "Is the 'Happiest Place on Earth' driving tigers and orangutans into extinction?"

Similarly, the threat of a boycott, or even the knowledge that an environmental group has organized successful campaigns in the past, may be enough to persuade a company that is concerned with its reputation to make concessions. One successful boycott that shows the potential of the strategy is RAN's effort to pressure Burger King to stop buying cattle that had been raised on land that had originally been tropical rainforest but had been cleared to provide pasture land. Initially, it was not clear whether RAN's campaign would succeed, but the boycott cut Burger King's sales and forced the company to cancel $35 million in beef contracts. Burger King began to buy beef exclusively from sources that did not cause global warming and took other steps to establish a proenvironment record.[13]

Effective media-oriented boycotts have the following characteristics in common. First, a well-known person or organization—that is, someone who can draw media attention—should announce the boycott. The strategy is so well known that the Environmental Change Institute at the University of Oxford conducted a study to identify the celebrities whom the public regards as most influential with respect to global warming.[14] These figures include Al Gore, Bill Clinton, Oprah Winfrey, Bono, Angelina Jolie, George Clooney, and Johnny Depp. When organizations can convince people of this caliber to be spokespeople for their campaigns, they are more likely to be effective.

The second point is related to the first: The announcements and boycott events should be as theatrical and attention grabbing as possible. Marches, stockholder meeting protests, and press conferences on oil-soaked beaches after oil spills are all tactics that have been used by boycott organizers. In recent years, organizations have turned to digital campaigns, in which they use a viral strategy to encourage people to share the organization's message with friends and family, as well as with their online social networks. In 2009, for example, the Reality Coalition—a joint project of the Alliance for Climate Protection, League of Conservation Voters, Natural Resources Defense Council, National Wildlife Federation, and Sierra Club—teamed up with award-winning filmmakers Joel and Ethan Coen to direct a new ad challenging the coal industry to debunk the myth of "clean coal." The video went viral in green communities. In one commentator's words, "News of this video spread faster than an Al Gore book launch."[15] Digital campaigns can also be powerful because they weaken the traditional gate-keeping role of mainstream media channels. As a result, campaigns that begin online often receive coverage in the mass media.

Third, the reasons for the boycott should be clearly explained and should seem legitimate to the public. If they seem to be valid reasons, a victory is far more likely. In their successful campaigns against Home Depot, Lowe's, and other users of wood and paper from old growth forests, RAN emphasized that the clearing of these forests is a huge source of greenhouse gas emissions that cause climate change. The argument that saving the forests would help save the planet resonated with consumers.

To make sense of the buycotting side of political consumerism, it is useful to distinguish between indirect and direct buycotts. Indirect buycotts refer to certifications by independent organizations that products meet certain political, health, or environmental standards. In contrast, direct buycotts are campaigns to target particular firms and encourage people to buy their products.

Independent nonprofit groups and government agencies run a number of indirect buycott programs to encourage people to buy particular products. These certification programs differ from commercial advertising, because their goals are not to make money. Instead, they want to encourage people to purchase environmentally sustainable products. In addition, they want to give companies incentives to be more environmentally sustainable by redesigning their products or changing their production practices.

The Green Seal program, for example, was founded in 1989 to promote environmentally sustainable products. In order to protect the environment and preserve natural resources, the organization develops science-based environmental standards for products ranging from paper and soap to hotels, and it allows companies to use the Green Seal logo on their products to show that they meet certain environmentally friendly criteria. It also posts information about products on its website for politically motivated consumers. Other certification programs include those of the US Forest Stewardship Council and the US Green Building Council, Fair Trade USA, the US Department of Agriculture's Organic Seal, and the US Environmental Protection Agency's ENERGY STAR ® program.

Political Consumerism and Climate Change

Perhaps the role of political consumerism in climate change politics can best be seen in a car—the Toyota Prius. The Prius is the first commercial hybrid gas–electric car in the world. It was introduced in Japan in 1997 and then sold in the United States starting in 2000. Although its career was bumpy in the early years, it became a huge success. By April 2011, one million cars had been sold in the United States.[16]

The Prius has been successful because it has come to symbolize the fight against climate change and therefore appeals to politically motivated consumers. The Prius is far more fuel efficient than its rivals with conventional internal combustion engines. Because of its hybrid engine, it gets 51 miles per gallon in city driving, far more than a similarly sized conventional car like the Honda Civic, which gets only 28 miles per gallon. This helps explain why people buy it despite the fact that it costs about $8,000 more than a comparable Honda Civic.

Of course, political consumerism is not the only reason the Prius has done so well. Studies of Prius owners identify five clusters of reasons people buy the cars.[17] First, people care about the environment and want to reduce their carbon footprints. Second, people like new technology and enjoy the idea of being on the leading edge of technological development. Third, people are concerned about complying with social norms. If they believe that they are living in communities that are proenvironment, purchasing a Prius shows that they are conforming to prevailing values. Fourth, some people believe that driving hybrids will help make the United States more energy independent, so we will not be subject to the ups and downs of international oil prices. Fifth, people want the financial and other material benefits they get from owning a Prius. Fuel efficiency saves owners money at the gas pump, and in some states Prius owners benefit from being allowed to drive in carpool lanes.

The motives of caring about the environment and about how people are viewed by their friends and neighbors are often found together in practice. When the *New York Times* asked Prius owners why they bought their cars, many responses were similar to that of a Philadelphia resident who said, "I really want people to know that I care about the environment. . . . I like that people stop and ask me how I like my car."[18] In short, owning a Prius makes a statement, and the desire to make that statement helps fight climate change. This is the flip side of the reputational motive for joining a boycott. Some people feel that violating a boycott will harm their personal reputations. Similarly, some people believe that driving a Prius or joining another buycotting campaign enhances their reputations.

Other energy-related purchasing decisions also affect the climate. The ENERGY STAR program, for example, may not have the splashy appeal of a Prius, but it has proven to be very effective. The program was launched in 1992 by the US Environmental Protection Agency and the US Department of Energy to reduce energy consumption and greenhouse gas emissions, and to save consumers money. The program labels products with the trademarked ENERGY STAR logo to help consumers identify energy efficient products. If a product such as a television meets the program's energy efficiency standards, the company making it can voluntarily choose to put the ENERGY STAR logo on it. Consumers who recognize the label will know that purchasing such a product will not only help the environment but will also save them money over time, because it uses less energy than comparable products without the logo.

One might think that the ENERGY STAR program works because people simply want to save money. That is clearly not the case. Early research after the first energy crisis of 1973–74 showed that relatively few people made the long-term energy- and money-saving decision to buy products that were more energy efficient just to save money. People discount future savings. One study, for example, found that people would pay only $2.12 to $2.45 for each dollar of annual energy savings when buying a refrigerator.[19] Yet refrigerators typically last 10 years or more, so rational people should be willing to pay far more in order to save money over the life of the appliance. Only when the ENERGY STAR program began putting out the message that saving energy was about more than saving money did energy efficient consumer behavior really start spreading. Slogans such as "Save energy, save money, protect the environment" work because they trigger political consumerism.[20]

The effectiveness of ENERGY STAR points to an important fact about political consumerism. People can be political consumers only if they are informed consumers. On some issues, the information demands are virtually zero. If someone wishes to be a vegetarian, it is easy to avoid beef and other products with meat. On other issues, organizations and companies invest a great deal of time and money to inform consumers. With respect to the Prius, for example, Toyota spent a great deal of money on advertising campaigns in which they highlighted the vehicle's fuel efficiency and hybrid engine. The news media gave the Prius a great deal of attention as well. Still other issues require consumers to identify and understand product labeling schemes. Without the ENERGY STAR labeling scheme, for example, few people would know which computers, printers, monitors, refrigerators, and other household appliances are the most energy efficient. That is why ENERGY STAR was set up, and it is why other programs—such as Green Seal, the Scientific Certification Systems' green product certifications, Green-e, the Federal Trade Commission's Energy Guide label, and the US General Services Administration's Carbon Footprint Tool—provide critical information to political consumers who care about climate change. Like other forms of political action, political consumerism requires information.

The success of the Prius and the ENERGY STAR program are great indicators of the importance of politics in the marketplace, but these examples cannot tell us how many people make purchasing decisions with environmental concerns such as climate change in mind. According to a series of national polls from the Yale Project on Climate Change Communication, however, a fair amount of consumers make purchasing decisions based on environmental concerns. In each of these polls, respondents were asked whether in the last year they had rewarded a company that was taking steps to reduce global warming by buying its products.[21] One-third said yes. Only a few, 8 percent, said they did it many times, but another 20 percent said that they did it at least a few times. These data do not show whether the purchases were large or small (e.g., a car or environmentally friendly laundry detergent), but they do indicate that the decisions were explicitly intended to affect global warming, even if in a very small way. To put this in context, more people said that they engaged in this form of political consumerism than participated in politics in any other way except voting.

Political organizations and activists also use boycotts to address issues related to climate change. Like buycotting, boycotting is more common than one might suspect. The 2012 poll from the Yale Project on Climate Change Communication, mentioned above, revealed that one-quarter of American adults said that in the previous year they had "punished companies that are opposing steps to reduce global warming by not buying their products." RAN, Greenpeace, ForestEthics, and other groups routinely turn to boycotting to put pressure on firms to change their practices. The Ethical Consumer and some other websites list many of these boycotts.[22] One successful boycott that shows the potential of the strategy is RAN's effort to pressure Burger King to stop buying cattle that had been raised on land that had been tropical rainforest, as mentioned above.

The Effectiveness of Political Consumerism

Political consumerism by itself will not save the world from climate change. That would require action on a far broader scale than anything that consumer behavior alone can do. Yet political consumerism clearly has already brought about some change, and it has the potential for an even larger impact. Perhaps the best way to think about it is to consider it as one element of many in a strategy for addressing climate change.

The success of the Prius was mentioned in the previous section. As a result, Toyota estimates, from the time they were introduced through 2011, the one million–plus Prius cars on the road saved drivers $2.19 billion in fuel costs and reduced CO_2 emissions by 12.4 million tons.[23] Beyond the Prius, the spread of hybrid cars—helped by political consumerism—could have a substantial impact. One study projects that moving to hybrid-electric cars "can reduce gasoline consumption and greenhouse gas emissions 30 to 50% with no change in vehicle class and hence no loss of jobs or compromise on safety or performance."[24] That would have a major impact on our total greenhouse gas emissions.

ENERGY STAR has also been successful. Guiding people toward energy efficient purchases and encouraging them to think of themselves as saving the environment cut an estimated $47 billion in energy costs and reduced greenhouse gas emissions by 82 million metric tons from 1992 through 2006.[25] From 2007 through 2015, the program is projected to save an additional $90 billion in energy costs and prevent 203 million metric tons of carbon-equivalent from being released into the atmosphere. To put it in perspective, that is a 3.3 percent reduction in the projected carbon emissions of the residential and commercial sectors. That may not sound like much, but for one small government program helping consumers to make their decisions wisely, it is pretty good.

Measuring the impact of other types of buycotting on greenhouse gas emissions is more difficult to do. Recent survey evidence shows that a high percentage of Americans engage in buycotting of some sort, but the data do not indicate exactly what purchasing decisions were made, nor do they offer any basis for estimating the reduction in greenhouse gas emissions the buycotting caused. There were certainly some effects, but the answers must wait for future research.

Measuring the impact of boycotts is also difficult to do. Their targets are typically businesses, which have an incentive not to reveal how much a boycott hurt their sales or harmed them in other ways, or what internal decisions they made in response to the boycott. Nevertheless, we can see cases in which political activists have successfully used boycotts to advance environmental causes that helped slow global warming.

The Rainforest Action Network began working to stop clear-cutting of rainforests to provide land for palm oil plantations in 2007. Three years later, they began targeting General Mills, which used palm oil from these plantations in Cheerios, Hamburger Helper, and other products. The Rainforest Action League's opening move was to unfurl a banner on the front lawn of General Mills's headquarters in Minneapolis that said, "Warning! General Mills Destroys Rainforests!" Nine months later, General Mills capitulated, announcing that it would stop buying palm oil from companies that were accused of destroying rainforests, and that it would buy all of its palm oil from "responsible and sustainable sources" within five years.[26]

The Stop Esso Campaign was launched in 2001 to pressure Esso (the brand name under which ExxonMobil sells gas in Europe), BP Amoco, Shell, and other oil companies to stop donating money to the Global Climate Coalition (GCC) and other climate change denial organizations. Bianca Jagger, one of the leaders of the effort, said, "This is a way to tell Esso that it's not right for them to be claiming that there is no connection between CO_2 emissions and climate change."[27] Although Exxon-Mobil continues to fund climate change denial groups, other major corporations withdrew from GCC, and it collapsed in 2002. The growing realization in the corporate world that climate change is real was a major cause of GCC's demise, but fears of boycotts directed at the businesses that funded GCC also played a key role.

ForestEthics, working with the Dogwood Alliance USA and other groups, starting a campaign of boycotts and publicity in the early 2000s to persuade the office supply industry to reduce the amount of timber it took from North America's boreal forests and to increase the amount of recycled paper it sold. Their goals were to preserve North American forests, and by doing so to preserve habitat and species diversity, and to help stabilize the climate. In 2003, Staples Office Supplies gave in to the demands of ForestEthics; a year later, Home Depot also agreed to change its paper supply practices. Both companies now get high grades from ForestEthics'

Green Grades Report Card, and both market themselves as environmentally friendly companies.

Although we cannot quantify the impact of environmental boycotts in terms of a reduction in tons of greenhouse gases emitted, as we can with the Prius and ENERGY STAR, we can nevertheless see that they are effective tools in some situations. In addition, the number of buycotters and boycotters are high in comparison with other forms of political participation, other than voting. Given these numbers and the poll's finding that slightly more than half of the respondents said that they intended to continue rewarding and punishing companies for their climate change behavior at the same level or even more often, one must conclude that political consumerism has the potential for a substantial impact on corporate behavior and possibly even on climate change.

Concluding Remarks

In this chapter, we introduced political consumerism as a political tool people can use to effect change. Throughout history, people have politicized the market when national governments did not or could not address their political and social grievances. As Friedman notes, boycotts have "been used more than any other organizational technique to promote and protect the rights of the powerless and disenfranchised segments of society."[28] Notable examples of boycotts that led to social and political change include in the Montgomery Bus Boycott in the 1950s and the United Farm Workers union boycotts of grapes and lettuce in the 1970s. More recently, boycotts have also forced companies to adopt more environmentally sustainable practices, as the examples of Home Depot, Lowe's, Burger King, and General Mills show.

Today, consumer boycotts are even more effective, because activist organizations can use digital media technologies to organize and maintain boycott and buycott campaigns and to generate news coverage in the mainstream news media. The advantages digital media afford organizations have not gone unnoticed. For example, a "Boycott BP" Facebook group appeared after the BP oil spill in the Gulf of Mexico in 2010. At the time of this writing, nearly 800,000 people had "liked" the group, signaling their views to their social networks. Boycotts threaten companies' sales and reputations, and business leaders take them seriously. Although we recognize the need for national, regional, and international governmental action on problems related to climate change, we also view political consumerism as a meaningful form of behavior in an increasingly globalized economy without sufficient regulatory and enforcement mechanisms in place.

Public opinion surveys show that a large percentage of Americans actively boycott or buycott for reasons related to climate change. More people engage in political consumerism intended to minimize global warming than in traditional campaigning such as working in campaigns or donating to candidates. Political consumerism on climate change is less obvious than campaigning; while campaigns are conducted in public, most acts of political consumerism are less obvious. Campaign advertisements on television, for example, are much easier to observe than someone's decision to buy a fuel-efficient car or avoid buying gasoline at an ExxonMobil station. Still, political consumerism is widespread, and when it is focused in organized campaigns, it has the potential to have a substantial effect on climate change.

Suggested Readings

Friedman, Monroe. *Consumer Boycotts: Effecting Change Through the Marketplace and the Media.* New York: Routledge, 1999.

Maniates, Michael. "Individualization: Plant a Tree, Buy a Bike, Save the World?" In *Confronting Consumption,* edited by Thomas Princen, Michael Maniates, and Ken Conca, 43–66. Cambridge, MA: MIT Press, 2002.

Micheletti, Michele. *Political Virtue and Shopping: Individuals, Consumerism, and Collective Action,* 2nd ed. New York: Palgrave Macmillan, 2010.

Micheletti, Michele, Andreas Follesdal, and Dietlind Stolle. *Politics, Products, and Markets: Exploring Political Consumerism Past and Present,* 2nd ed. New Brunswick, NJ: Transaction, 2006.

Stolle, Dietlind, Marc Hooghe, and Michele Micheletti. "Politics in the Supermarket: Political Consumerism as a Form of Political Participation." *International Political Science Review* 26, no. 3 (2005): 425–269.

References

Andersen, Jorgen Goul, and Mette Tobiasen. "Who Are These Political Consumers Anyway? Survey Evidence from Denmark." In *Politics, Products, and Markets: Exploring Political Consumerism Past and Present,* 2nd ed., edited by Michelle Micheletti, Andreas Follesdal, and Dietlind Stolle, 203–222. New Brunswick, NJ: Transaction, 2006.

Baek, Young Min. "To Buy or Not to Buy: Who Are Political Consumers? What Do They Think and How Do They Participate?" *Political Studies* 58, no. 5 (2010): 1065–1086.

Baron, David P., and Daniel Diermeier. "Strategic Activism and Nonmarket Strategy." *Journal of Economics and Management Strategy* 16 (2007): 599–634.

Bennett, W. Lance. "Branded Political Communication: Lifestyle Politics, Logo Campaigns, and the Rise of Global Citizenship." In *Politics, Products, and Markets: Exploring Political Consumerism Past and Present,* 2nd ed., edited by Michelle Micheletti, Andreas Follesdal, and Dietlind Stolle, 101–125. New Brunswick, NJ: Transaction, 2006.

Brody, Jane. "Concern for Rain Forest Has Begun to Blossom." *New York Times,* October 13, 1987.

Copeland, Lauren. "Conceptualizing Political Consumerism: How Citizenship Norms Differentiate Boycotting From Buycotting." *Political Studies,* advance online publication (2013). doi:10.1111/1467-9248.12067.

Darnall, Nicole, Ceys Pointing, and Diego Vazquez-Brust. "Why Consumers Buy Green." In *Green-Growth: Managing the Transition to Sustainable Capitalism,* edited by B. Vazquez-Brust and J. Sarkis, 287–308. New York: Springer, 2012.

Environmental Change Institute, University of Oxford. *Climate Change & Influential Spokespeople—A Global Nielsen Online Survey.* June 2007. http://www.eci.ox.ac.uk/publications/downloads/070709nielsen-celeb-report.pdf.

ForestEthics. "Our Proudest Achievements." http://forestethics.org/proudest-achievements.

Friedman, Monroe. *Consumer Boycotts: Effecting Change Through the Marketplace and the Media.* New York: Routledge, 1999.

Friedman, Thomas. "A Tiger by the Tail." *New York Times,* June 1, 2001.

Hayes, Denis. *The Official Earth Day Guide to Planet Repair.* Washington, DC: Island Press, 1993.

Inglehart, Ronald. *Modernization and Postmodernization: Cultural, Economic, and Political Change in 43 Societies.* Princeton, NJ: Princeton University Press, 1997.

Koch, Wendy. "General Mills Boycotts Palm Oil That Destroys Rain Forests." *USA Today,* September 24, 2010. http://content.usatoday.com/communities/greenhouse/post/2010/09/general-mills-palm-oil-rainforest-destruction/1#.UOoRxax0mmR.

Leiserowitz, Anthony, Edward Maibach, Connie Roser-Renouf, Geoff Feinberg, and Peter Howe. *Americans' Actions to Limit Global Warming in September 2012.* Yale University and George Mason University. New Haven, CT: Yale Project on Climate Change Communication, 2012.

Maniates, Michael. "Individualization: Plant a Tree, Buy a Bike, Save the World?" In *Confronting Consumption,* edited by Thomas Princen, Michael Maniates, and Ken Conca, 43–66. Cambridge, MA: MIT Press, 2002.

Maynard, Micheline. "Toyota Hybrid Makes a Statement, and That Sells." *New York Times,* July 4, 2007.

Micheletti, Michele, Andreas Follesdal, and Dietlind Stolle. *Politics, Products, and Markets: Exploring Political Consumerism Past and Present,* 2nd ed. New Brunswick, NJ: Transaction, 2006.

Newman, Benjamin J., and Brandon L. Bartels. "Politics at the Checkout Line: Explaining Political Consumerism in the United States." *Political Research Quarterly,* advance online publication (2010). doi:10.1177/1065912910379232.

Rainforest Action Network. "Our Mission and History." http://ran.org/our-mission.

Romm, Joseph. "The Car and Fuel of the Future." *Energy Policy* 34 (2006): 2609–2614.

Sanchez, Marla C., Richard E. Brown, Carrie Webber, and Gregory K. Homan. "Savings Estimates for the United States Environmental Protection Agency's ENERGY STAR Voluntary Product Labeling Program." *Energy Policy* 36 (2008): 2098–2108.

Schwartz, Ariel. "FedEx, Office Depot Top ForestEthics' Green Grade Report Cards." *Fastcompany,* September 9, 2010. http://www.fastcompany.com/1688149/fedex-office-depot-top-forestethics-green-grades-report-card.

Strømsnes, Kristin. "Political Consumerism: A Substitute for or Supplement to Conventional Political Participation?" *Journal of Civil Society* 5, no. 3 (2009): 303–314.

Szasz, Andrew. *Shopping Our Way to Safety: How We Changed From Protecting the Environment to Protecting Ourselves.* Minneapolis: University of Minnesota Press, 2007.

Toyota USA Newsroom. "Toyota Sells One-Millionth Prius in the U.S." April 6, 2011. http://pressroom.toyota.com/article_display.cfm?article_id=2959&view_id=35924.

Ward, David O., Christopher D. Clark, Kimberly L. Jensen, Steven T. Yen, and Clifford S. Russell. "Factors Influencing Willingness to Pay for the ENERGY STAR Label." *Energy Policy* 39 (2011): 1450–1458.

Notes

* This material is based upon work supported by the National Science Foundation (NSF) under cooperative agreements #SES-0938099 and SES-0531184 to the Center for Nanotechnology in Society at the University of California Santa Barbara. Any opinions, findings, and conclusions or recommendations expressed in this material are those of the authors and do not necessarily reflect the views of the NSF.

1. Information retrieved from the organization's website: https://carrotmob.org/.
2. Lauren Copeland, "Conceptualizing Political Consumerism: How Citizenship Norms Differentiate Boycotting from Buycotting," *Political Studies*, advance online publication, doi:10.1111/1467-9248.12067.
3. Ronald Inglehart, *Modernization and Postmodernization: Cultural, Economic, and Political Change in 43 Societies* (Princeton, NJ: Princeton University Press, 1997); Lauren Copeland, "Value Change and Political Action: Postmaterialism, Political Consumerism, and Political Participation," *American Politics Research* (2013), advance online publication, doi:10.1177/1532673X13494235.
4. Jorgen Goul Andersen and Mette Tobiasen, "Who Are These Political Consumers Anyway? Survey Evidence From Denmark," in *Politics, Products, and Markets: Exploring Political Consumerism Past and Present,* 2nd ed., ed. Michelle Micheletti, Andreas Follesdal, and Dietlind Stolle (New Brunswick, NJ: Transaction, 2006), 205.
5. Original survey data were collected in the United States between December 7 and 21, 2011, by YouGov. The sample includes 2,200 US adults. See Copeland, "Conceptualizing Political Consumerism."
6. Although European studies of political consumerism find that political consumers are more likely to be female.
7. Benjamin J. Newman and Brandon L. Bartels, "Politics at the Checkout Line: Explaining Political Consumerism in the United States," *Political Research Quarterly* 64 (2011): 807.
8. Michael Maniates, "Individualization: Plant a Tree, Buy a Bike, Save the World?" in *Confronting Consumption,* ed. Thomas Princen, Michael Maniates, and Ken Conca (Cambridge, MA: MIT Press, 2002), 51.
9. Kristin Strømsnes, "Political Consumerism: A Substitute for or Supplement to Conventional Political Participation?" *Journal of Civil Society* 5, no. 3 (2009): 303–314.
10. Monroe Friedman, *Consumer Boycotts: Effecting Change Through the Marketplace and the Media* (New York: Routledge, 1999), ch. 2.
11. W. Lance Bennett, "Branded Political Communication: Lifestyle Politics, Logo Campaigns, and the Rise of Global Citizenship," in *Politics, Products, and Markets,* ed. Micheletti, Follesdal, and Stolle, 112.

12. Rainforest Action Network, "Disney and RAN Agree to Historic Commitment For Indonesia's Forests," http://ran.org/disney-and-ran-agree-historic-commitment-indonesia%E2%80%99s-forests.

13. Rainforest Action Network, "Our Mission and History," http://ran.org/our-mission; Jane Brody, "Concern for Rain Forest Has Begun to Blossom," *New York Times,* October 13, 1987.

14. Environmental Change Institute, University of Oxford, *Climate Change & Influential Spokespeople—A Global Nielsen Online Survey,* June 2007, http://www.eci.ox.ac.uk/publications/downloads/070709nielsen-celeb-report.pdf.

15. Brian Merchant, "The Top 9 Green Viral Videos (and One You Should Never Have to See, Ever)," http://planetgreen.discovery.com/work-connect/top-viral-videos.html.

16. Toyota USA Newsroom, "Toyota Sells One-Millionth Prius in the U.S.," April 6, 2011, http://pressroom.toyota.com/article_display.cfm?article_id=2959&view_id=35924.

17. Nicole Darnall, Ceys Pointing, and Diego Vazquez-Brust, "Why Consumers Buy Green," in *Green-Growth: Managing the Transition to Sustainable Capitalism,* ed. B. Vazquez-Brust and J. Sarkis (New York: Springer, 2012), 287–308.

18. Micheline Maynard, "Toyota Hybrid Makes a Statement, and That Sells," *New York Times,* July 4, 2007.

19. David O. Ward et al., "Factors Influencing Willingness to Pay for the ENERGY STAR Label," *Energy Policy* 39 (2011): 1450–1458.

20. Ibid.

21. Anthony Leiserowitz et al., *Americans' Actions to Limit Global Warming in September 2012,* Yale University and George Mason University (New Haven, CT: Yale Project on Climate Change Communication, 2012), 27.

22. See http://www.ethicalconsumer.org/boycotts/currentboycottslist.aspx and http://www.fuw.edu.pl/~pmh/boycott.html.

23. Toyota USA Newsroom, "Toyota Sells One-Millionth Prius."

24. Joseph Romm, "The Car and Fuel of the Future," *Energy Policy* 34 (2006): 2609.

25. Marla C. Sanchez et al., "Savings Estimates for the United States Environmental Protection Agency's ENERGY STAR Voluntary Product Labeling Program," *Energy Policy* 36 (2008): 2098–2108.

26. Wendy Koch, "General Mills Boycotts Palm Oil That Destroys Rain Forests," *USA Today,* September 24, 2010; Rainforest Action Network, "General Mills Takes Bold Steps Away From Palm Oil Controversy," http://ran.org/general-mills-takes-bold-steps-away-palm-oil-controversy#.

27. Thomas Friedman, "A Tiger by the Tail," *New York Times,* June 1, 2001.

28. Friedman, *Consumer Boycotts,* 3.

CHAPTER 9

The Politics of Urgent Transition

Thomas Princen

FOR A THEORY OF WORLD POLITICS to be more than explanatory, to be predictive and, given a set of values such as sustainability and justice, normative, it must do more than assume a trajectory from the past into the future. Extrapolation may have been a reasonable default for the future when history seemed to repeat itself, when statistical reasoning created a bias to the status quo, when rationalist thinking lent to incremental change, and when economic policy tended to the cautious midrange of growth (not so slow as to raise unemployment, not so fast as to ignite inflation). But today we operate in a "world risk society," where risks cannot be limited in time or place, entail little accountability, and cannot be compensated or insured against.[1] What is more, abundant evidence points to discontinuous change, indeed to change that challenges the very foundations of the modern world. This is a society without precedent in human history and thus requires new approaches to the future.

The approach I take in this chapter is to start with the biophysical and build to the ethical, primarily in the context of energy in general and fossil fuels in particular, and with a process of social change I call *localization*. With evidence of scientific and experiential of fundamental shifts in climate, water supplies, soil fertility, and a host of other ecological features, a parallel shift in how societies organize themselves to adapt will, I assume, be occurring. I call this *transition* to parallel usage in the fields of energy and technology. What's more, because so many of these shifts occur in open, nonlinear systems with limited knowability and predictability (e.g., the climate), they create what are best called "super wicked problems": Time to avert dangerous, even catastrophic outcomes is running out (biophysical irreversibilities and social path dependencies are real); those who seek solutions also cause the problem (we all use fossil fuels); requisite authority to act is weak or nonexistent (no global institution exists for climate change, biodiversity loss, etc.); conventional policy processes (e.g., cost-benefit analysis) discount the future irrationally.[2]

Anticipating and preparing for such shifts may well be defining conditions of the politics of the coming years and decades. Doing so rapidly, resisting the temptation to presume the biophysical conditions of the past (that human activity is minor relative to resources and waste sinks, that "new sources" will replace fossil fuels, that technologies will enable business-as-usual, only cleaner and greener), delaying because there is no apparent crisis, points to a politics of urgency. Those politics have at least two dimensions: one, a rejection of fossil fuels and, two, an embrace of localizing trends.

So in this chapter I characterize the politics of transition, of fundamental shift, here in the 21st century. It is necessarily broad brush and future oriented. And it is normative: I presume that, as crises multiply and publics everywhere recognize that the future will have unprecedented challenges, these publics will desire a transition that is peaceful, just, and sustainable.

The questions, then, are (again, starting with the biophysical and moving to the ethical):

1. What are the political implications of continued fossil fuel use when a scientific consensus (along with increasing instances of personal experience—think heat waves and hurricanes) points to catastrophic loss on the current path? Volumes have been written on managerialist approaches, yet precious little on the need to "go to the source"—to the fossil fuel industry, private and governmental, and its financial and political enablers. Hence I develop a politics, both economic and ethical, of the rejection of fossil fuels.

2. When modern societies wean themselves off fossil fuels, whether through deliberate action or because physical conditions compel it, how will societies reorganize? Here I posit a process of social change called *localization,* one with local, national, and international dimensions.

First, though, I define transition and set the context.

Transition

I take transition to be a long-term process of social change, one that connects a previous state to a new state with little or no chance of returning to the previous state and that provides a distinct context for critical decision making.[3] Thus, a wood-powered industry transitions to coal powered after timber supplies are exhausted and infrastructure for coal is locked in. A plantation economy built on slave labor transitions to a market economy and full human rights when constitutions and treaties are written to ban slave trading and ownership. Small-holder farming transitions to industrial farming when public policies favor large-scale operations, and, once the capital investments are made, farmers can't go back.

Transition differs from emergency and crisis in the time frame and in expectations of the future. An *emergency* is immediate, with rapid response designed to restore life to what it was before the emergency. A flood prompts evacuation and rescue and, with cleanup and rebuilding, people get back to normal. A *crisis* spans months and years, resisting clean resolution but, with correctives and fine-tuning, is eventually overcome, and business-as-usual returns. The oil shocks of the 1970s created an economic crisis among oil-importing countries. With national strategic reserves and new sources, importers resumed economic growth.

Transition, by contrast, spans long periods of time, decades or centuries, and leaders and publics alike see no possibility of going back. Industrialization, a process of technological, energetic, and cultural change, spanned several centuries, and few can seriously predict a return to subsistence agriculture and craft manufacturing.

In both emergency and crisis, the drivers of social change remain largely the same. For example, technologies develop incrementally, markets continue to expand, representative democracy spreads, and cultural exchange continues apace. In transition, the drivers change fundamentally. A revolutionary technology emerges (e.g., the steam engine, the computer), a critical resource is exhausted (e.g., timber, soil), a command economy collapses, a state balkanizes, entire populations migrate or become isolated. The change, once again, can span decades, even centuries.

Because the difference between emergency and crisis, on the one hand, and transition, on the other, is in the drivers of social change and in the time frame, the politics and decision making differ from one to the other. A full-blown theory of transition would spell out that decision making in terms of motivation, perception, cognition, worldview, and ethics. Lacking such a theory, here, suffice it to say, transition is not a simple extrapolation of emergency and crisis decision making, both of which are fields of study and practice that are well developed. In emergency and crisis a primary objective is to "get back to normal," whereas in transition it is to "get to a new normal." Decisions in emergency and crisis are reaction, response, recovery; in transition they are adaptation, innovation, entrepreneurship.

Finally, transition implies some degree of inexorability. Forces are underway, whether human caused or not, that are largely out of the control of conventional, consequential decision-making bodies (from families to the international community). While many emergencies can be prevented (with, e.g., smoke alarms, traffic signals, tornado warnings) and crises forestalled (with, e.g., strategic petroleum reserves, stock market shut-down mechanisms), transitions entail large, whole-system change.

In this chapter, then, I explore transition and the politics thereof first by positing a plausible scenario for the future: cheap energy (that is, energy that has been cheap economically, energetically, and environmentally) is coming to an end, and payments for past environmental abuses are coming due. I then develop likely decision-making dynamics, primarily at the global level. The exercise, it should be noted, is necessarily future oriented (into the ecologically and geologically far future) and highly uncertain, what naturally characterizes all but historical transitions.

Biophysical and Industrial Context

The fossil fuel era, that period when fossil fuels have dominated fuels from all other sources, began in the United States and worldwide only in the 1890s.[4] Until recently, that era has been defined by two unassailable facts—cheap, high-density energy has been easy to get; and it has been available in ever-increasing amounts. To put numbers to these two facts, humans have extracted and burned roughly a trillion barrels of conventional oil to date. There's another trillion available. And there's some 4 to 5 trillion, some say as much as 18 trillion, barrel equivalents in other fossil fuels.[5] If the first trillion has been enough to change the climate and alter ocean chemistry, possibly irretrievably and with huge human costs, then adding the emissions of another trillion or more will compound those effects (not just add to them or continue them) and will almost certainly be catastrophic.

That, in a nutshell, is the biophysical situation, the material context that will shape politics of all sorts in the coming decades. But this situation did not "just happen." A complex set of decisions by powerful actors, reinforced by mass publics (primarily as consumers) created it. So the industrial context of the fossil fuel era begins with key actors and their sources of influence.

One measure of the fossil fuel industry's influence is the fact that 88 percent of the world's energy comes from fossil fuels.[6] Sixty-one percent of that is produced by national oil companies created, subsidized, and defended by national governments.[7] Another measure is that the petroleum industry is the world's largest, capitalized at $2.3 trillion and comprising 14.2 percent of all commodity trade.[8] What's more, it is by far the most capital-intensive industry—$3.2 million is invested for every person employed. By comparison, the textile industry is capitalized at $13,000, the computer industry at $100,000, and the chemical industry at $200,000 per worker.[9] And the petroleum industry is among the most profitable. In 2008, for example, ExxonMobil made $11.68 billion in second-quarter profits, amounting to profits of some $1,400 per second, and ranking 45th on a list of the top hundred economic entities in the world, a list that includes national governments.[10]

Perhaps the industry's greatest source of influence, though, is its ability to advance a vision, one of abundant and cheap energy, of powering and defending nations, of feeding and sheltering billions of people. It is a vision with appeal to nearly every sector of a modern industrial society—manufacturers, investors, military and political leaders, consumers.

Emissions Management

For all the concern about fossil fuels, including their dominance of political economies worldwide, the predominant approach to both ground-level air pollution and high-level climate change is to *manage emissions,* to reduce a couple centuries of history to one chemical element, carbon, to, in effect, seek end-of-pipe solutions when the real problem is upstream, in a global infrastructure and power structure that is extremely adept at laying new pipes. In the context of climate change, the implicit theory of social change goes something like this:

> Climate change is a global problem. Like global security and global trade, it must be managed globally. Global managers are of two sorts—those with the authority, states, and those with the requisite knowledge, scientists. Only the global elites can work at this level. Only the scientists can observe the problem (through their instruments), requiring as it does vast data sets, sophisticated modeling, and the funds to support the science. Only the managers can marshal the resources to tackle such a gargantuan problem. Only the diplomats can reach the agreements that overcome the global collective action problem. Only the policy makers can arrange incentives, so their respective publics behave correctly.

Curiously missing from this global management formulation of the problem and the rightful actors who would solve it are those actors who organize to pull fossil fuels out of the ground—oil, gas, and coal companies, both private and state, and the industrial development arms of governments. Missing are the complex networks of actors who accomplish the remarkable transformation of raw materials to usable products (shippers, refiners, manufacturers, distributors, petrochemical companies, etc.), who ensure the flow of such materials (domestic security and court systems, international security forces), and who finance it all (bankers, investors, consumers). My impression is that nearly all research funding for climate change goes to "the science," some to the economics and intergovernmental relations, but next to nothing to an understanding of the political economy of extraction and combustion.[11] In all these, "the problem" is construed as emissions.

From a transition perspective (as opposed to the crisis perspective, often implicit, of global management), the central problem is not about what is done *after* extraction and combustion; it is, in the first instance, about extraction itself. Put differently, regarding climate change, it is not about "carbon" but about "fossil fuels."

A carbon focus is reductionist, possibly the greatest and most dangerous reductionism of all time: A 150-year history of complex geologic, political, economic, and military security issues all reduced to one element—carbon. This chemical framing implies that the problem arises after a chemical transformation, after fuels are burned. It effectively absolves of responsibility all those who organize to extract and process and distribute, state and nonstate. It puts the burden on governments and consumers to rectify the situation, a situation that is otherwise presumed to be given—normal or inevitable or desirable. Finally, "carbon" portrays the global ecological predicament as completely one dimensional: Solve the climate change problem—that is, deal with carbon—and everything else, from toxics to particulates to weather extremes, follows. In practice, when "carbon" is replaced with biofuels, for example, what follows is eroded soil, clear-cut forests, and, quite literally, starving people and riots.

To focus on *fossil fuels,* by contrast, is to ask about the status of oil and coal and natural gas in the ground and how it is and why it is that these complex hydrocarbons come out of the ground. It does not take such how-and-why questions as self-evident (people want the energy, producers get it). A fossil fuel focus directs attention, analytic and eventually political attention, upstream to a whole set of decisions and incentives and structures that conspire to bring to the surface hydrocarbons that otherwise sit safely and permanently in the ground. It forces one to consider that, once fossil fuels are extracted, their by-products—ground-level pollution and atmospheric greenhouse gasses—inevitably and unavoidably move into people's bloodstream, into ecosystems, and into the atmosphere and oceans.

Global management schemes may derive their legitimacy from their very rationality and scientific soundness (including economic calculations) but their appeal—to environmentalists and oil companies alike it seems—derives from a different source: They are essentially *a*political; they don't attempt to rewrite the constitutional rules of the game—namely, that extraction proceeds full speed ahead. Theirs is not a politics of urgency, of fundamental transition. It is a politics of accommodation. Instead of confronting the power of the fossil fuel complex, the global managers create their own politics, a tame politics, a collaborative, rationalist, scientific, managerial politics, one that ruffles few feathers but keeps everyone pointed in the same direction. Here the tug-and-pull is over data and modeling; the give-and-take is about using carbon judiciously, like a commodity; the distributing is about putting carbon in the right place; the influencing is about educating so everyone understands the science and the costs. The game is one of haggling over prices and pricing, of finding places to put the offending stuff, of rationalizing cleanup.

The politics is, notably, of a wholly different sort from the politics of the fossil fuel complex itself, where extractive industries largely write the rules of the game and decide who gets access to the resources. The politics of the extractive industries, from the earliest days of underground coal mining and land-based oil drilling, is indeed the politics of extraction, total extraction.[12] For that politics, everything else—distributing carbon credits, assigning liabilities for oil spills—is child's play, a convenient distraction, a great diversion.[13] From a fossil fuel transition perspective, a normative shift as monumental as any in human history will define the politics of the coming decades.[14]

For such a politics, the current state of affairs can be summed up thus: If fossil fuel use could be presumed net beneficial in its first century, it cannot in its second and third centuries. Current technologies, market demand, and geopolitical strategic imperatives are sufficient to drive the extraction and burning of catastrophic amounts of fossil fuels. Given this and the state of local emissions control and global management schemes to date, as noted, it is reasonable to assume that when fossil fuels are extracted, their by-products *will* enter people's bloodstreams and water supplies, the oceans and the atmosphere. Consequently, a normative shift on the order of abolition, industrialization, democratization, international peace, and suffrage will be needed for an early exit. Moral entrepreneurs will have to find leverage points in current material systems, even if they cannot imagine, let alone offer a plan for, the post–fossil fuel era.[15]

So the difference, both analytic and rhetorical, between "carbon" and "fossil fuels" is the difference between reductionism and complex systems, between global management and deliberative decision making, between management and elimination, between end-of-pipe and prevention, between cleanup and abstention, between technocracy and democracy. It is between getting back to normal after an emergency or crisis versus transitioning to a new normal, one without the dominance of fossil fuels and the fossil fuel industry.

Fossil Fuel Rejection

So if the predominant approach to ground-level air pollution, high-level climate change, persistent toxic substances, and a host of other environmental ills is to manage emissions, how could one go beyond end-of-pipe, beyond crisis management, to fundamental transition?

First, consider the implications of end-of-pipe. Economically, end-of-pipe effectively says the problem occurs at the end of a long chain of production and consumption decisions. Pollutants emerge after the goods are produced, as a by-product, as an unfortunate yet unavoidable side effect. All previous steps in the production-consumption-disposal chain are indeed about "goods," that which can be presumed beneficial or benign. Exploration, testing, drilling, transporting, securing, processing, manufacturing, distributing, advertising, lobbying are all given—given, that is, by a combination of entrepreneurial spirit, extraordinary risk taking, technological innovation, capital investment, managerial choice, and, once all this is instituted, consumer demand and political imperative. So construed, emissions are merely an unfortunate and inconvenient side effect. But because the goods come from all that is given and captured by those actors so engaged, and the bads come from the emissions that nobody wants, a society (read government or taxpayers) is obligated to ameliorate the bads. This division of goods and bads is terribly convenient for those who actually make the key decisions and reap so much of the rewards. It is an ethic in its own right, one that says producing goods is inherently good. But in the larger scheme of things, in that system that incorporates both the extraction *and* the disposal over geologically long periods of time—the only system that could legitimately be called an "ecosystem"—it is hardly ethical when downstream vulnerable populations now and in the future face the bads.

Second, from a political economy perspective, what we know to be the natural order of things is actually deliberately constructed, an order whose rules have evolved over time to suit well those who benefit most. Certainly all players—rule makers and rule followers—benefit in some fashion (investors gain returns on their investments, risk-taking employees are paid well, consumers have cheap and abundant energy), but such benefits are time constrained, as are all mining operations. Boom times may last years, even a century or two, but the bust is perfectly predictable in a mining economy. The fossil fuel binge may be hugely profitable and amazingly stimulating, not to mention convenient and fun; it may transport, heat, and feed billions, but it is *not sustainable*. It will end. The only question is how: how peacefully and democratically, how planfully and equitably.

All this leads to the uncomfortable conclusion (for fossil fuel proponents anyway) that the only realistic means of stopping fossil fuel emissions is to leave fossil fuels in the ground, not to stop cold their use, but accept the urgency and start stopping. The only safe place for fossil fuels is in place, where they lie, where they are solid or liquid (or, for natural gas, geologically well contained already), where their chemistry is mostly of complex chains, not simple molecules like CO_2 that find their way out of the tiniest crevices, that lubricate tectonic plates perpetually under stress, that react readily with water to acidify the oceans and that float into high places filtering and reflecting sunlight, heating beyond livability the habitats below. This approach, I assert, is realistic in both a biophysical *and* a political sense.[16]

If rejecting fossil fuels is ultimately a moral question, then the ultimate strategy for bringing the fossil fuel era to an early end may well be *delegitimization*. By delegitimization I do not mean a vilification of the fossil fuel industry. The industry has already had a century and more of vilification, starting with charges against Rockefeller's Standard Oil (the "Octopus") and continuing through to today. (A former Shell president titled his book *Why We Hate the Oil Companies*.) Nor do I mean a repudiation of the industry's antidemocratic, antienvironmental tactics. Rather, by delegitimization I mean the reconceptualization and *revalorization of fossil fuels* or, to be precise, humans' *relations* with fossil fuels. I mean a shift from fossil fuels as constructive substance to fossil fuels as destructive substance, from necessity to indulgence, from sustenance to addiction, from a "good" to a "bad," from lifeblood (of modern society) to poison (of a potentially sustainable society). I mean a shift from fossil fuels as that which is normal to that which is abnormal.

In other words, fossil fuels will make a *moral transition* in parallel to the *material transition*. Much as slavery went from universal institution to universal abomination and as tobacco went from medicinal and cool to lethal and disgusting, the delegitimization of fossil fuels will flip the valence of these otherwise wondrous, free-for-the-taking complex hydrocarbons. And rather than pin blame on "big bad oil (and coal) companies" or, even worse, on "all of us" because we all use fossil fuels, delegitimization simply recognizes that a substance once deemed net beneficial can become net detrimental. All it takes is a bit of evidence (in the case of fossil fuels, a mountain of evidence already exists), some incisive critics, effective communication, and, for the "moral entrepreneurs," a whole lot of persistence and willingness to themselves be vilified.[17] It would start with the simple observation that there are some things humans can not handle. Their level of understanding, their susceptibility to convenience or power, and their inability to organize globally and for the long term all mitigate against having such things as ozone-depleting substances, lead, drift nets, land mines, rhino horns, and, someday perhaps, nuclear weapons and nuclear power plants. This, then, would be the essential politics of rejecting fossil fuels.

A perceptual shift appears to have already begun, as suggested by Dennis Chong earlier in this volume, establishing a necessary precursor to a political transition. That shift is, for instance, from environment as amenity to environment as existential threat, from fossil fuels as the essential lifeblood to fossil fuels as a global threat to life as we know it. Those who first made the perceptual shift were climate scientists and their followers in the academic, policy, and activist communities, always a tiny minority, yet at times vocal and influential. Then the general public began to shift, prompting a backlash. Perhaps the most notable shift, however, has been among those who, irrespective of ideology or religious belief, are making changes *in practice*. Water masters, forest and range managers, irrigators, farmers, timber companies, fishermen, landscapers, weather forecasters all are changing the way they do things. Even oil companies that explore, prepare sites, drill and pump with time horizons of decades must, to protect investments and deliver a product, factor in climate change (not to mention political change). For these actors, to be effective, indeed profitable, is to plan for long-term changes in water availability, temperature, pests, storms, ice, sea level, and so forth. It is simply bad policy and bad business to do otherwise.

These shifts—among scientists, policy elites, resource managers, business people—are arguably pragmatic, not moral. The source of the moral shift is likely to come initially from those who suffer the most from such changes, those who will be driven from their ancestral lands onto marginal lands or into urban slums. As these peoples—not all from the Global South—gain a voice, they may well drive the moral shift. In fact, they arguably are increasingly doing so, just out of view of mainstream analysts. "Any lingering postcolonial dismissal of environmentalism as marginal to 'real' politics," writes literary scholar Rob Nixon, "is belied by the proliferation of indigenous environmental movements across the global South . . . locally motivated, locally led, and internationally inflected."[18] While much of this politics is aimed at oppressive regimes and rapacious corporations, a logical material focal point is fossil fuels.[19]

Not so long ago, people North and South had little reason to believe that oil wealth brought anything but great prosperity to their localities and countries. But contrary experience is widespread and growing, from the Niger Delta to Louisiana's chemical alley, from rig- and mine-worker deaths to automobile deaths. What is more, a scholarly literature on the so-called resource curse for many oil-producing countries is extensive, albeit still contested.[20] The industry may still have great power and a compelling vision, but the social and economic costs throw into question the long-standing presumption of net beneficence. "The irony of oil wealth," writes political scientist Michael Ross, is that "the greater a country's need for additional income—because it is poor and has a weak economy—the more likely its oil wealth will be misused or squandered. . . . Since the oil nationalizations of the 1970s, the oil-producing countries have had less democracy, fewer opportunities for women, more frequent civil wars, and more volatile economic growth than the rest of the world, especially in the developing world." In addition, Ross finds, "By 2005, at least half of the OPEC countries were poorer than they had been thirty years earlier."[21] From a national security perspective, Jim Woolsey, former director of the CIA, says that, from his perspective, "It was obvious that oil was dominant in a lot of places that generated trouble. There's almost nothing that doesn't get better if you move away from dependence on oil."[22] Even industry insiders have taken stock and are trying to imagine a different world. "The resources are there," writes John Hofmeister, former president of the Shell Oil Company US. "The question is: do we *want* to continue to use these fossil fuels at current—or increasing—rates until they are eventually exhausted? The answer, unequivocably, is no. The economic, social, and environmental costs of such an approach are becoming ever clearer and ever higher."[23] Or, as the German Advisory Council on Global Change put it, "The 'fossil-nuclear metabolism' of the industrialized society has no future. The longer we cling to it, the higher the prices will be for future generations."[24] In short, for all the promises of prosperity and all the power of fossil fuel actors, the deliberate construction of fossil fuel's net beneficence and inevitable use is beginning to crumble.

Thus a growing delegitimization of fossil fuels is a realistic scenario, not as a response to the abstract claims of the climate science community, but as a response to existential threats perceived by peoples everywhere (some of which are indeed climate induced), even some in the fossil fuel industry. What the climate scientists and others started, yet can not finish with their top-down, expert-led, managerialist schemes and technological fixes, will be augmented and accelerated by moral commitments. While fossil fuel–dependent societies cannot stop cold, from a transition perspective, their overarching policy must be to *start stopping now.*[25]

To delegitimize a substance (or a process like exploring and drilling), as opposed to condemning an actor or charging a system, puts the focus on the offending substance or, more specifically, on its *use.* Fossil fuels are perfectly "natural"; ancient uses were, for all I can tell, benign. In a strategy of delegitimization, the burden shifts from the contest of interest groups (e.g., enviros versus industrialists) to a contest over the politics of the good life. Industrialists have enacted one vision of the good life. Its efficacy in the 20th century can be debated (and probably will be for the rest of the 21st century), but the politics of delegitimization are about now and the future, including the distant future. What is more, as a normative theorizing exercise, it is about creating the good, the good *given* the biophysical trends underway.

So a politics of fossil fuel delegitimization, of deliberately accelerating a society's withdrawal from oil, gas, and coal dependence, ahead of the geologic imperative, ahead even of the economic and financial imperatives, is ultimately an ethical act. It puts front and center the "harm to others" criterion and relegates to the wings the economic and political (as in electoral and legislative politics) criteria. What is more, a politics (as in the shaping of society's core values and steering a particular path) of fossil fuel exit is one of temporal extension, of taking seriously humans' past and their future, including their geologically and ecologically distant past and future. Temporal extension necessitates ethical extension—from present generations to past and future generations, from us to other, from human life to nonhuman life, from resources to ecosystems, from extraction to regeneration, from material gain to societal integrity and spiritual uplift, from goods-are-good-and-more-goods-must-be-better to the "good life." A politics of fossil fuel exit is thus more than effecting the next energy transition, more than arresting climate change, more than shifting to a postindustrial world. It is a moral confrontation with a wildly successful material order, an order that has heretofore been presumed net beneficial, salutary, indeed essential and just. So construed, a politics of delegitimizing becomes *a politics of creating*—let's create an economy without fossil fuel dependency. This stands in contrast to the prevailing politics, that of crisis management and marginal improvement—let's control the ill effects but don't dare question the emergency response and crisis management.

How might a politics of creating occur? If a globalized order depended on cheap energy (cheap economically, energetically, and environmentally), might the new order be a localized order? It is to this question that I now turn.

Localization: From Distance to Place

Localization can be seen as a logical outcome of the disappearance of a one-time, historically tiny period—that is, the age of abundant, highly available energy.[26] How localization proceeds thus becomes one of the defining questions of a politics of transition. So what is localization? Or, what could it be, given a set of trends, especially the accelerated end of the fossil fuel era?

First, an analogy from physics. Add energy to a system—a solution of salts, say, or a herd of elephants, or an economy—and things move. And with more and more energy, they move faster and farther. At some point, things disperse—liquids boil, animals migrate, people travel, goods get shipped.[27]

The infusion of ever-increasing energy could be a model of globalization. Cheap and abundant energy creates the centrifugal forces that make the world go around, that is, the contemporary world of global commerce and politics. And an economy so energized must somehow dissipate the energy. So people travel farther and faster, goods cross continents and oceans, raw materials flow to their highest return on investment, finished products to their highest demand. Among the economic processes that result from (and drive) such *distancing* are commercialization, commodification, consumerism, mechanization, and specialization.[28]

With more and more energy, the system eventually breaks. The beaker dries to a crusty residue of salts and then cracks; the economy's bubbles burst; its excesses bite back. This, arguably, is the stage of the experiment global society is now experiencing. The appropriate response for some is the techno-fix. Business-as-usual, greened up and made cost-effective, is the order of the day. "The answer" is biofuels, hydrogen, fuel cells, more nuclear power plants, hyperefficient cars, superefficient light bulbs, better batteries, and, possibly the mother of all gambles, geoengineering—sequestering emitted carbon in the ground, manipulating the climate.

For others, though, the response is localization. That is, as less energy becomes available, or as societies deliberately limit extraction and slow their fossil fuel combustion, or as financial systems collapse, societies will necessarily and unavoidably become more localized. The end of cheap and abundant energy means the end of distancing, the end of a bottle of water from France, a strawberry in winter, a two-hour commute by car, a quick trip to London, a 300-horsepower truck to haul groceries, a night sky that's never dark, a heating and cooling system that runs 24/7/365.

So what is localization? What can it be, good and bad? What *should* it be given the social goals of a peaceful, democratic, just, and ecologically sustainable transition out of the hyperenergetic era? I discuss these questions both descriptively and normatively, identifying drivers of localization and, having started with the biophysical, finishing with a brief consideration of the ethical.

Not the Local, Not Globalization Reversed

First, localization is not "the local." Local is any geographically bounded activity or cultural pattern that is less than the regional or national or global, geographically less, that is. A Walmart store is local: The building and inventory and employees are entirely within one locality. Much of its water and energy is supplied locally. In a different way, my locally owned, locally operated Ann Arbor delicatessen, Zingerman's, is, by design, local—no franchises, no national expansions; its owners have made the commitment to stay local. They've also made the commitment to stocking the best cheeses and wines and chocolates, wherever they find them—Europe, New Zealand, Africa.

So, a given enterprise, however global (e.g., Walmart) or local (e.g., Zingerman's) it may appear, is simultaneously local and global. In practice, it actually constitutes itself at a scale greater than the geographically local. Localization, then, is not simply that which occurs at the local level. Nor is it maintaining the local or improving the local, however progressive or green or economically stimulating that may be. Abundant literatures exist on "the local," whether the emphasis is on the city, the small town, or that ever-contested and utterly nebulous notion, the "community." For present purposes, I will assume these literatures cover their ground well—that is, they lay out what needs to be done to strengthen a locality, whether that strengthening be economic, cultural (including "lifestyle"), environmental, or with respect to social justice. But they do not confront dramatic drops in energy and material availability. They do not predict the strains and opportunities that will arise with the end of cheap oil, the mitigation of and adaptation to climate change, the drawdown of fresh water, and so forth. They do not, in short, chart a course for the *transition*.

Localization also is not simply the opposite of globalization. Antisweatshop activism can result in labor standards, inspections, and certification. These measures, organized globally, are a countervailing force to globalization. They adjust the dominant trajectory, and arguably only the magnitude, not the direction. It's still globalization. Localization, by contrast, charts a fundamentally different course, a trajectory that no amount of fine tuning of globalization can achieve.[29]

Localization as a Politics

Localization, then, is neither "the local" nor the opposite of globalization. So what is its politics? In a nutshell, it is a process of social change that points toward localities, yet is not strictly about localities. As overextended economies and ecosystems spend themselves, economic and cultural exchange will shift inward, moving toward a new equilibrium state, however dynamic, where societies live on current sunlight, on recharging water, and on self-renewing soils and biota. The centrifugal forces of globalization—cheap energy, technological advance, intensified commercialization and communication, displaced wastes, concentrated economic and political power—shift to the centripetal forces of diminished energy and material, enhanced personal security, and declining faith in centralized authority. Localization entails increased attention to the tangible, the interpersonal, the face-to-face, and to a place-based community. It entails more direct connection to the natural world, especially that which provides sustenance—food and shelter, for instance, and that which enhances well-being—greenery and clean rivers, for instance.[30] This is not to say that localization is inattentive to the regional, national, and international. Quite the contrary. It is only that the primary *direction*, the *focus of attention*, is inward, to the concrete, to direct interactions with social and natural systems, rather than outward, to the abstractions and diversions of economic and technological systems. To localize, in short, is to shift from worldviews that separate and distance to those that connect and bring close.

Toward an Ethics of Localization

Because localization is a social process driven by material reality (the decline of cheap energy, the rising costs of environmental abuses) and arguably by declining trust in global, centralized institutions, its norms must have both biophysical and social dimensions.

Regarding the biophysical, Leopold's land ethic may be sufficient for the end point of localization.[31] That is, once societies have adapted to (or at least have accepted the reality of) the end of cheap energy, the impossibility of endless resource depletion, and the necessity of curbing climate disrupting emissions, the land ethic—supplemented by a couple of decades of sustainability science and philosophy—can probably serve well enough as a guide.[32]

The localization imperative, though, is to guide the *transition,* to make the downshift proceed in ways that are peaceful, democratic, just, and ecologically sustainable. Arguably, these four criteria are not mere ideals. They are necessary conditions for adapting to inexorable external shifts. For instance, a violent transition is likely to destroy the requisite human relations, the social capital, and the biophysical requisites of human security, personal well-being, and social integrity in transition or in a steady state.

Another ethical consideration is the public's reaction to the unavoidable reduction in consumption of energy and material. One reaction is likely to be "Woe is us": Civilization is coming to an end, we're doomed, time to turn off the lights, crawl into the cave, and shiver in the dark. Countless book and movie titles play on this sentiment. The other is "Good! It's about time!": We've been overconsuming for too long, obsessing with shopping and fashion and status symbols and throwaway consumables.

The woe-is-us reaction derives from a materialist ethic that says happiness (or pleasure or satisfaction or utility maximization or well-being) comes from what we own, what we purchase after earning income. Materialists identify "well-being with buying, having, and displaying consumer goods, especially those that bring comfort and convenience," writes philosopher David Crocker.[33] The more goods and services and the more consumer choices the better. And better for the individual and society, because, after all, society is the aggregation of all individuals. This materialist and individualist ethic underlies neoclassical economics and the entire neoliberal, highly distanced, expansionist order.

The "good-it's-about-time" reaction derives from an antimaterialist ethic. From the earliest days of the free market system, critics have charged that consumption is unduly focused on things, if not downright corrupting. Well-being comes from renouncing commodities, forsaking worldly possessions, freeing oneself of material attachments. In less extreme forms, antimaterialists emphasize "inner rationality, self-possession, and self-sufficiency," writes Crocker, and the "fulfillment that comes through personal relationships."

But as Crocker points out, these two ethics are, in a sense, of a sort, each depending on the other. "Although materialism and anti-materialism both contain some truth, they are also guilty of exaggeration. Those who endorse one of these views are typically engaged in an overreaction against the other, while others find themselves torn between the two."[34] The dichotomy is false and, for Crocker's purpose of constructing a consumption norm and mine of constructing a transition norm, must be transcended. One means, Crocker argues, drawing on Amartya Sen who in turn draws on many others all the way back to Aristotle, is to focus on the sources and meanings of *well-being,* in particular, to focus on what Sen calls "capabilities." It is not hard to imagine the policy response would be of two sorts:

1. avoiding a reduction in consumption, denying that the downshift is necessary or inevitable (here is where the techno-green, no-sacrifice, grow-our-way-out with renewables and efficiencies comes in), and

2. celebrating the simple, even ascetic lifestyle while condemning the McMansion dwellers and SUV drivers.

The avoidance response is well underway and needs little elaboration. The celebration response is likely to lead to approaches that are insular, either individualistic or oriented to geographically constrained entities—"communities." They neglect the societal, the national, and international. At best it is a strategy of social change based on aggregation: As more people discover the blessings of simple living and the value of community, society as a whole changes. At worst, it is a strategy doomed to survivalist outposts and walled enclaves.

The materialist/antimaterialist frame is thus not helpful in conceptualizing a positive localization, that is, in constructing an ethic of localization. Crocker argues that a capabilities approach transcends these two sides of the same materialist coin. A capabilities approach is not a compromise between the materialist and the antimaterialist. It is an approach grounded in a psychologically and philosophically rich notion of well-being.

For present purposes, a critical feature of the capabilities approach is the observation that well-being is not consumption dependent. That is, above some basic level of material subsistence (the actual level highly culturally determined) and probably below some level of affluence (the rich, having systematically deskilled themselves, thus may be more vulnerable to downshift than others), well-being depends mostly on *competencies,* on the ability to solve problems and produce useful and pleasing things. Well-being does not depend on a flow of goods and services, let alone a continuously increasing flow of goods and services.[35]

In sum, localization, whether positive or negative, is likely to occur as cheap and abundant energy declines and trust in centralized institutions further erodes. Attention shifts to the near-at-hand, the tangible, the face-to-face. And it does so with particular concern for "the basics"—food and water, shelter, and transportation. Positive localization will focus more on competencies and well-being than on goods and consumption.

Transition Politics

I have used the term *transition* to imply social change more fundamental, possibly more disruptive, than that presumed in conventional international relations (with the possible exception of nuclear exchange). And I have argued that two defining, baseline features of the associated politics will be the rejection of fossil fuels and the embrace of localizing tendencies. These features are baseline in two senses. One is biophysical: All economies rest on natural systems—water and soil, minerals and petroleum, forests and grasslands, estuaries and reefs, the atmosphere and oceans. What is more, they ultimately rest on the continued functioning and ecological integrity of such systems. The other is socioeconomic: All political economies, from craft to market driven, from heavy industry to high technology, from mom-and-pop retail to high finance—all such economies rest on the agricultural. No matter how sophisticated the technologies or how complex the financial instruments, no economy can function without a steady supply of food (and fiber and building material). Put differently, whereas in "normal times" the economic super-structure garners the bulk of public attention (output, employment, interest rates), what underpins it is the economic infrastructure (farms, local retail, start-ups). In times of dramatic change, both emergency (flood, tornado, drought) and crisis (climate change, national debt, persistent civil strife), attention shifts to this infrastructure (ensure water and food supplies, heating and cooling facilities). The politics of this shift, when fundamentals are in jeopardy and delay is hugely risky, are *the politics of urgent transition*.

A focus on transition, then, is motivated in part by the observation that whereas fundamental issues could be ignored in the past, they cannot now. Repeated emergencies and crises portend a wholly different sort of change than that which commands political attention when abundant energy and food supplies and a strong agricultural sector (indeed a growing one) could be assumed. Those "20th-century" conditions no longer apply. The evidence for resource depletion and waste sink filling is overwhelming. And whereas analysts—mostly technical, mostly from the fields of engineering and economics and related professions and discipline—are well aware of the so-called energy transition (as if it were strictly a physical matter), few are working on the political transition (aside from calls for policy based on "sound science" and for more "political will").

No substances better defined the material conditions of the 20th century than fossil fuels. Since their rise and dominance as a power source and as essential strategic commodity, they have been presumed net beneficial, right to be used, wrong to be left in the ground. Yet like so many substances and practices through modern times, where benefits are immediate and concentrated among the few and costs delayed and dispersed to the many, there comes a time when citizens and leaders ask whether fossil fuels are compatible with the good life, indeed, with life as we know it. The early abolitionists posed similar questions, and after decades of dogged research, public education, and politicking, convinced their society that the answer with respect to slavery was no. If there was a good life under the institution of slavery, it was for the few (traders, planters, investors), and everyone else, from Europe to Africa to the Americas, was degraded, free and slave alike. Now such questions are being posed with respect to nuclear weaponry and nuclear power. Along the way, international society has concluded that biological weapons and land mines, ozone-depleting substances, mile-long driftnets, and a dozen or so persistent organic pollutants are too incompatible with the good life—that is, the good life for all, for all time.

To engage a 21st-century politics of fossil fuels, one of accelerated phaseout, is to open a public debate heretofore unimaginable. It would be a debate that goes to the core of modern, industrial, expansionist and consumerist society. It would challenge the prerogatives of the most powerful actors in the world, state and nonstate. It would question globalization on fundamental energetic and ecological grounds and entertain an entirely different process of social change, localization. Not to open such a debate would be to abrogate responsibility for existing wrongs visited upon select populations (e.g., polar peoples, rig workers, miners) and to avoid responsibility for coming wrongs that, if worst-case scenarios play out, would dwarf all previous human-caused calamities.

The behaviors most in demand in this transition are likely to be those of anticipation, innovation, and entrepreneurship; the politics those of urgency and adaptation. Underpinning all that is an ethic of the sustainable use of resources and competency among citizens. All these are worthy topics for further research in a politics of urgent transition.

Suggested Readings

De Young, Raymond, and Thomas Princen. *The Localization Reader: Adaptations for the Coming Downshift.* Cambridge, MA: MIT Press, 2012.

Nixon, Rob. *Slow Violence and the Environmentalism of the Poor.* Cambridge, MA: Harvard University Press, 2011.

Princen, Thomas, Jack P. Manno, and Pamela Martin. "Keep Them in the Ground: Ending the Fossil Fuel Era." In *State of the World 2013: Is Sustainability Still Possible?*, edited by The Worldwatch Institute, 161–171. Washington, DC: Island Press, 2013.

Smil, Vaclav. *Energy Transitions: History, Requirements, Prospects.* Santa Barbara, CA: Praeger, 2011.

Swilling, Marck, and Eve Annecke. *Just Transitions: Explorations of Sustainability in an Unfair World.* Tokyo: United Nations University Press, 2012.

References

Beck, Ulrich. *World at Risk.* Cambridge, UK: Polity, 2009.

Berglof, Annie Maccoby. "At Home: Jim Woolsey. The Former Head of the CIA Wants to Wean the US Off Oil." *Financial Times,* July 6, 2012.

Brandt, Adam R., and Alexander E. Farrell. "Scraping the Bottom of the Barrel: Greenhouse Gas Emission Consequences of a Transition to Low-Quality and Synthetic Petroleum Resources." *Climatic Change* 84, no. 3–4 (2007): 241–263.

Crocker, David A., and Toby Linden. *Ethics of Consumption: The Good Life, Justice, and Global Stewardship.* Lanham, MD: Rowman and Littlefield, 1998.

Day, Jr., John W., Charles A. Hall, Alejandro Yáñez-Arancibia, David Pimentel, Carles Ibáñez Martí, and William J. Mitsch. "Ecology in Times of Scarcity." *BioScience* 59 (2009): 321–331.

De Young, Raymond, and Thomas Princen. *The Localization Reader: Adapting to the Coming Downshift.* Cambridge, MA: MIT Press, 2012.

El-Gamal, Mahmoud A., and Amy Jaffe. *Oil, Dollars, Debt, and Crises: The Global Curse of Black Gold.* Cambridge, UK: Cambridge University Press, 2010.

Hall, Charles, Pradeep Tharakan, John Hallock, Cutler Cleveland, and Michael Jefferson. "Hydrocarbons and the Evolution of Human Culture." *Nature* 426 (2003): 318–322.

Hofmeister, John. *Why We Hate the Oil Companies: Straight Talk From an Energy Insider.* New York: Palgrave Macmillan, 2010.

Humphreys, Macartan, Jeffrey Sachs, and Joseph E. Stiglitz. *Escaping the Resource Curse.* New York: Columbia University Press, 2007.

Karl, Terry Lynn. *The Paradox of Plenty: Oil Booms and Petro-States.* Berkeley: University of California Press, 1997.

Kasser, Tim. *The High Price of Materialism.* Cambridge, MA: MIT Press, 2003.

Klare, Michael T. *Resource Wars: The New Landscape of Global Conflict.* New York: Henry Holt, 2002.

Levin, Kelly, Benjamin Cashore, Steven Bernstein, and Graeme Auld. "Overcoming the Tragedy of Super Wicked Problems: Constraining Our Future Selves to Ameliorate Global Climate Change." *Policy Science* 45 (2012): 123–152.

Nadelmann, Ethan A. "Global Prohibition Regimes: The Evolution of Norms in International Society." *International Organization* 44, no. 4 (1990): 479–526.

Nixon, Rob. *Slow Violence and the Environmentalism of the Poor.* Cambridge, MA: Harvard University Press, 2011.

Princen, Thomas. "A Sustainability Ethic." In *Handbook of Global Environmental Politics*, edited by Peter Dauvergne, 466–479. Cheltenham, UK: Edward Elgar, 2012.

Princen, Thomas, Michael Maniates, and Ken Conca. "Distancing: Consumption and the Severing of Feedback." In *Confronting Consumption,* edited by Thomas Princen, Michael Maniates, and Ken Conca, 103–131. Cambridge, MA: MIT Press, 2002.

Princen, Thomas, Jack P. Manno, and Pamela Martin. "Keep Them in the Ground: Ending the Fossil Fuel Era." In *State of the World 2013: Is Sustainability Still Possible?,* edited by The Worldwatch Institute, 161–171. Washington, DC: Island Press, 2013.

Ross, Michael Lewin. *The Oil Curse: How Petroleum Wealth Shapes the Development of Nations.* Princeton, NJ: Princeton University Press, 2012.

Schellnhuber, Hans Joachim, Dirk Messner, Claus Leggewie, Reinhold Leinfelder, Nebojsa Nakicenovic, Stefan Rahmstorf, Sabine Schlacke, Jürgen Schmid, and Renate Schubert. *World in Transition: A Social Contract for Sustainability.* Berlin: German Advisory Council on Global Change, 2011.

Smil, Vaclav. *Energy Transitions: History, Requirements, Prospects.* Santa Barbara, CA: Praeger, 2010.

Smil, Vaclav. "Global Energy: The Latest Infatuations." *American Scientist* 99, no. 3 (2011): 212–219.

Swilling, Mark, and Eve Annecke. *Just Transitions: Explorations of Sustainability in an Unfair World.* Tokyo: United Nations University Press, 2012.

Notes

1. Ulrich Beck, *World at Risk,* trans. Ciaran Cronin (Cambridge, UK: Polity, 2009).
2. Kelly Levin, Benjamin Cashore, Steven Bernstein, and Graeme Auld, "Overcoming the Tragedy of Super Wicked Problems: Constraining Our Future Selves to Ameliorate Global Climate Change," *Policy Science* 45 (2012): 123–152.
3. Even an energy transition is not strictly a physical shift, from wood to coal, say. It is at once a technological and cultural shift, from low-intensity to high-intensity firing, from a decentralized, locally inflected economy to a centralized, nationally and internationally inflected economy. Defining transition in decision-making terms thus stresses the role of choice and the conditions under which those choices are made. I thank Raymond De Young for working through this definition with me.
4. Vaclav Smil, *Energy Transitions: History, Requirements, Prospects* (Santa Barbara, CA: Praeger, 2011, figures on 63, 108.
5. A. R. Brandt and A. E. Farrell, "Scraping the Bottom of the Barrel: Greenhouse Gas Emission Consequences of a Transition to Low-Quality and Synthetic Petroleum Resources," *Climate Change,* 84 (2007): 241–263; C. Hall, P. Tharakan, J. Hallock, C. Cleveland, and M. Jefferson, "Hydrocarbons and the Evolution of Human Culture," *Nature* 426 (2003): 318–322; International Energy Agency, *World Energy Outlook 2010* (Paris: OECD/IEA, 2010).
6. Vaclav Smil, "Global Energy: The Latest Infatuations," *American Scientist* 99 (2011): 212–219, quote on 217.
7. David G. Victor, David R. Hultz, and Mark C. Thurber, "Introduction and Overview," in *Oil and Governance: State-Owned Enterprises and the World Energy Supply,* ed. David G. Victor, David R. Hultz, and Mark C. Thurber (Cambridge, UK: Cambridge University Press, 2012), 3–31, data on 3.
8. Michael L. Ross, *The Oil Curse: How Petroleum Wealth Shapes the Development of Nations* (Princeton, NJ: Princeton University Press, 2012), 3.

9. Ibid., 45.

10. Steve Coll, *Private Empire: ExxonMobil and American Power* (New York: Penguin Press, 2012), 8.

11. For a concise and insightful critique of the "scientific separatism" and "technological determinism" that underlies this theory of social change, see Sheila Jasanoff, "The Essential Parallel Between Science and Democracy," *D.C. Science,* February 17, 2009, http://seed magazine.com. See also Bruno Latour, *Politics of Nature: How to Bring the Sciences Into Democracy,* trans. Catherine Porter (Cambridge, MA: Harvard University Press, 2004).

12. For such history, see Barbara Freese, *Coal: A Human History* (New York: Penguin, 2003); Daniel Yergin, *The Prize: The Epic Quest for Oil, Money, and Power* (New York: Simon and Schuster, 1991); Steve Coll, *Private Empire: ExxonMobil and American Power* (New York: Penguin, 2012). For the ethics of total extraction, see Thomas Princen, "A Sustainability Ethic," in *The Handbook of Global Environmental Politics,* ed. Peter Dauvergne (Cheltenham, UK: Edward Elgar, 2012), 466–479.

13. To be clear, the "extremely powerful actors" are not just the oil companies and petro states. They are that larger edifice with a revolving door in the middle, the one where the players are extractors and financiers one moment, rule makers the next, and then back again.

14. For readers still resistant to the idea of futuring and normative theorizing, consider that this is precisely what the foreign policy, security, and business communities, both academic and practitioner, routinely do. See also Levin et al, "Overcoming the Tragedy of Super Wicked Problems."

15. On moral entrepreneurs, see Ethan A. Nadelmann, "Global Prohibition Regimes: The Evolution of Norms in International Society," *International Organization* 44, no. 4 (1990): 479–526.

16. For an extended treatment of the politics of deliberately keeping fossil fuels in the ground, with case studies, see Thomas Princen, Jack P. Manno, and Pamela Martin, "Keep Them in the Ground: Ending the Fossil Fuel Era," in *State of the World 2013: Is Sustainability Still Possible?,* ed. The Worldwatch Institute (Washington, DC: Island Press, 2013), 161–171; and Thomas Princen, Jack P. Manno, and Pamela Martin, eds., *Keep 'Em in the Ground: Ending the Fossil Fuel Era,* under review.

17. Nadelmann, "Global Prohibition Regimes."

18. Rob Nixon, *Slow Violence and the Environmentalism of the Poor* (Cambridge, MA: Harvard University Press, 2011), 255.

19. For evidence of that rising voice, see Nixon, *Slow Violence;* Jerry Mander and Victoria Tauli-Corpuz, eds., *Paradigm Wars: Indigenous Peoples' Resistance to Globalization* (San Francisco: Sierra Club Books, 2006); Jack Manno and Pamela Martin, "The Influence of Yasuní and Haudenosaunee Thought on the Global Movement for the Rights of Nature, the Rights of Indigenous Peoples and Humanity's Responsibility to 'Leave it in the Ground,'" and Robin Broad and John Cavanagh, "Keep-It-In-The-Ground— Beyond Fossil Fuels & Climate Change: The Case of Gold Mining in El Salvador," both in *Keep 'Em in the Ground,* ed. Princen, Manno, and Martin.

20. Mahmoud A. El-Gamal and Amy Myers Jaffe, *Oil, Dollars, Debt, and Crises: The Global Curse of Black Gold* (Cambridge, UK: Cambridge University Press, 2010); M. Humphreys, J. Sachs, and J. Stiglitz, eds., *Escaping the Resource Curse* (New York: Columbia University Press, 2007); Terry Lynn Karl, *The Paradox of Plenty: Oil Booms and Petro-States* (Berkeley: University of California Press, 1997); Michael

Klare, *Resource Wars: The New Landscape of Global Conflict* (New York: Holt, 2002); Ross, *The Oil Curse.*

21. Ross, *The Oil Curse,* 236, 253, 189.

22. Annie Maccoby Berglof, "At Home: Jim Woolsey. The Former Head of the CIA Wants to Wean the US Off Oil," *Financial Times,* July 6, 2012.

23. John Hofmeister, *Why We Hate the Oil Companies: Straight Talk From an Energy Insider* (New York: Palgrave Macmillan, 2010), 48.

24. German Advisory Council on Global Change, "World in Transition: A Social Contract for Sustainability: Summary for Policy-Makers" (Berlin: German Advisory Council on Global Change, 2011), 1–26, quote on 25.

25. For elaboration of this argument and consideration of the ethics of fossil fuel use, past and future, see Thomas Princen, "Start Stopping: The Ethics of Leaving Fossil Fuels in the Ground," in Princen, Manno, and Martin, *Keep 'Em in the Ground.*

26. On an energy-scarce future and coming transition, see John W. Day Jr., Charles A. Hall, Alejandro Yáñez-Arancibia, David Pimentel, Carles Ibáñez Martí, and William J. Mitsch, "Ecology in Times of Scarcity," *Bioscience* 59 (2009): 321–331.

27. I thank Patrick Murphy of Community Solutions, Yellow Springs, Ohio, for this analogy. Personal communication, November, 2007.

28. For an ecological economics of distancing, see Thomas Princen, "Distancing: Consumption and the Severing of Feedback," in *Confronting Consumption,* ed. Thomas Princen, Michael Maniates, and Ken Conca (Cambridge, MA: MIT Press, 2002), 105–131.

29. For elaboration of the localizing features of transition, see Raymond De Young and Thomas Princen, *The Localization Reader: Adaptations for the Coming Downshift* (Cambridge, MA: MIT Press, 2012).

30. For data on such trends, especially in a rural context, see Thomas A. Lyson, *Civic Agriculture: Reconnecting Farm, Food, and Community* (Medford, MA: Tufts University Press, 2004); and, from an urban perspective, see Rachel M. Krause, "Climate Policy in U.S. Cities," this volume, Chapter 4.

31. By "end point" I do not mean to imply that there will be a single equilibrium state once all the flux of the transition is ironed out. Among other things, the ideals of democracy and sustainability require never-ending vigilance and adaptation. Thus, a "dynamic equilibrium" is more what I have in mind with this usage of end point.

32. Aldo Leopold, "The Conservation Ethic," in *The River of the Mother of God and Other Essays by Aldo Leopold,* ed. Susan L. Flader and J. Baird Callicott (Madison: University of Wisconsin Press, 1933), 181–192.

33. David A. Crocker, "Consumption, Well-Being, and Capability," in David A. Crocker and Toby Linden, eds., *Ethics of Consumption: The Good Life, Justice, and Global Stewardship* (Lanham, MD: Rowman and Littlefield, 1998) 366–390, quote on 368.

34. Ibid., 370–371.

35. Tim Kasser, *The High Price of Materialism* (Cambridge, MA: MIT Press, 2002).

Index

Note: Page numbers in *italics* indicate figures and tables.

Abbott, Greg, 61, 64–65
ACEEE (American Council for an Energy-Efficient Economy), 67–68, *163,* 164
Actions, in context of attitudes about climate change, 120–121
Age cohort differences (generational change), and public opinion on climate change, 116–118, *117, 118*
Alliance for Climate Progress, 154–155
Alliance of Small Island States (AOSIS), 20–21
Alliance to Save Energy, *163,* 164
American Clean Energy and Security Act in 2009, 48, 58, 63, 74, 135
American Clean Energy and Security Act of 2009, 39
American Council for an Energy–Efficient Economy (ACEEE), 67–68, *163,* 164
American Electric Power Co. v. Connecticut in 2009, 42
American National Election Study (ANES), 173
American Recovery and Reinvestment Act in 2009, 39
ANES (American National Election Study), 173
AOSIS (Alliance of Small Island States), 20–21
Apollo Alliance, *151,* 155–156, 158, *159,* 160, *162*

Arizona, and Proposition 202 or CGMI (Citizen's Growth Management Initiative), 176–177
ARs (Assessment Reports), 3, 5, 12, 14, 16–17, *17,* 19, 28n8
Assembly Bill 32 (California's 2006 Global Warming Solutions Act), 66–67
Assessment Reports (ARs), 3, 5, 12, 14, 16–17, *17,* 19, 28n8
Asymmetry, and international climate policy making as collective action problem, 13–15

Bäckstrand, K., 152, 153, 155
Ballot measures, and states' policy making. *See also* Electoral participation, and consumer political action; States' climate policies
overview of, 24–25, 171–172, 190–191
California's Alternative Fuels Initiative or Proposition 10 and, 186–187
case studies of renewable energy measures, 185–189
causal factors for proenvironmental successes and, 180–181, *182,* 183–185, *184*
Colorado's Renewable Energy Act or Amendment 37 and, 185–186, 188
Maine's Energy Efficiency Bonds Issue or Question 2 and, 187–188

Michigan's renewable energy amendment
or Proposal 3 and, 188–189
proenvironmental failures and, 172–177
proenvironmental successes and, 172–181,
179, 180, 182, 183–185, *184*
public opinion on climate change policy
and, 172–173
Barnes, Kay, 93
Bartels, Brandon L., 203
Benedick, Richard E., 136
Benegal, Salil, 127
Berk, Richard, 121
Berry, Jeffrey M., 102
Betsill, Michele, 87
Biophysical (human activity) context of
energy, and politics of transition, 221
Bloomberg, Michael, 101
BlueGreen Alliance, *151,* 154–156,
159–160, *162*
Bornstein, Nicholas, 177
Bowler, Shaun, 177, 181, 183
Boycotts, and consumer political action,
198–200, *199,* 211
Boykoff, Jules M., 124
Boykoff, Maxwell T., 124, 130
British Petroleum (BP), 135, 200,
206, 212–213
Brulle, Robert J., 126
Bulkeley, Harriet, 87
Bush, George H. W., 57
Bush, George W., 6, 37–38, 44, 48,
60, 70, 112, 125–126
Buycotts, and consumer political action,
198–200, *199, 201,* 208–211
Byrd-Hagel Resolution, 28n12, 34

California
Air Resources Board of, 77
Assembly Bill 32 or Global Warming
Solutions Act in 2006, 66, 67
climate policies in, 65–67
Proposition 9 or Clean Air Initiative, 176
Proposition 10 or Alternative Fuels
Initiative, 186–187
Proposition 20 or Coastal Zone
Conservation Act, 176

Proposition 39 or Clean Energy
Jobs Act, 171
Proposition 50, 175
Proposition 65 or toxics initiative, 176
Senate Bill 375, 82
Cannon, Jonathan Z., 44
CAPs (climate action plans), 88, 90–94
Carbon dioxide (CO_2) 35–37, 42, 44–46,
52n14, 53n42, 60, 74, 82, 115, 131,
146, 148, 205. *See also* CO_2 emissions
Carbon dioxide equivalent (CO_2e) gases
data on, 100, 101–102
Carmichael, Jason, 126
Carrotmob, 197–198
Carter, Jimmy, 146
Case studies, of renewable energy measures,
185–189
Catholic Coalition on Climate Change, *151,*
158, *159, 162*
Causal factors, for proenvironmental
successes, 180–181, *182,* 183–185, *184*
CCP (Cities for Climate Protection), 85, *86*
CDM (clean development mechanism), 6
Certification programs, and consumer
political action, 208, 210
CFCs (chlorofluorocarbons), 3, 28n9, 134
C40 Cities Climate Leadership group, 103
Chesapeake, *151,* 158, *159, 162*
Chicago Department of Environment
(DOE), 98
Chicago, Illinois CCAP (climate action
plan), 98–99, *99*
China, 6, *7,* 8–9, 14–15, 20, 35–36, *36,* 44
Chlorofluorocarbons (CFCs), 3, 28n9, 134
Chong, Dennis, 176–177, 183, 187, 226
Cities' climate policy making
overview of, 23–24, 82–84, 102–103
CAPs and, 88, 90–94
CCP and, 85, *86*
C40 Cities Climate Leadership group
and, 103
Chicago, Illinois CCAP and, 98–99, *99*
CO_2e gases data and, 100, 101–102
explicit climate actions and, 91–92, *92*
GHG and, 82–83, 88–89, *89, 90,*
91–92, *92–93*

government-focused initiatives in context of community emissions and, 90–91

historical context for, 84–88

ICLEI and, 83, 85, *86,* 87–88, 93, 103

implicit climate actions and, 91–92, *92*

Kansas City, Missouri (KCMO) climate protection movement and, 93–94

Kyoto Protocol and, 86

local climate policies and, 88–94, *89, 90*

locally led initiatives' impacts on, 100–102

Mayor Climate Protection Agreement and, 86, *86*

motivations for, 94–99, *95, 97*

New York City climate protection program and, 101–102

Sierra Club's Cool Cities program and, 103

Urban C02 Reduction Project and, 85

U.S. Conference of Mayors' Climate Protection center and, 103

Cities for Climate Protection (CCP), 85, *86*

Citizen's Growth Management Initiative (CGMI or Arizona's Proposition 202), 176–177

Civic environmentalism coalitions, 153–156, *162*

Civic participation, and consumer political action, 205, *205*

Civic society, and climate change. *See* Ballot measures, and states' policy making; Consumer political action, and climate change; National climate change movement in US; Politics of transition, and climate making policy; Public opinion on climate change

Clean Air Act, 23, 33–34, *40,* 42–45, 42–46, 43–46, 48–49, 60–62, 111

Clean development mechanism (CDM), 6

Clean Energy Works, *151,* 155, *159,* 160, *162*

Clean Water Act, 46, 50

Climate action plans (CAPs), 64, 87–89, *90,* 91–94, 98

Climate Council, 134

Climate policies. *See* Cities' climate policy making; Courts, and national climate policy making; National climate policy making; States' climate policies

Climate SOS, *151,* 155, *159, 162*

Climate Stewardship Act of 2003, 37–38

Clinton, Bill, 34–35, 173, 207

CO_2e (carbon dioxide equivalent) gases data on, *7,* 100, 101–102

CO_2 emissions

data on, 1, 6–8, *7,* 28n1, 35–36, *36*

international climate policy making as collective action problem and, 14, 16, 18–19

international climate regime and, 1, 6–10, 28n1

national climate policy making and, 35–36, *36,* 52n18, 53n42

public opinion on climate change and, 111–112, 119, 127

rainforest's protection and, 205

See also Carbon dioxide (CO_2)

Coalition formation, and national climate change movement in US, 149, 151

Coal phaseout, and states' climate policies, 71–72

Coen, Ethan, 207

Coen, Joel, 207

Colorado, and Amendment 37 or Renewable Energy Act, 184–186, 188

Conference of the Parties (COP), 5, 149. *See also* COP (Conference of the Parties), COP-7 in Marrakesh, COP-13 in Bali, COP-15 in Copenhagen, COP-16 in Cancun, COP-17 in 2017 in Durban, COP-17 in Durban, COP-18 in Doha, COP-19 in Warsaw

Conflicting opinions, on climate change, 111–112

Congressional hearings, on climate change, 148–149, 151

Connecticut v. American Electric Power Co. in 2005, 42, 43

Consensus, on climate change, 111–112

Conservative movement against scientific
consensus, and public opinion on
climate change, 112, 115–116,
125–126
Consumer political action, and climate
change
overview of, 25, 197–198
boycotts and, 198–200, *199*, 211
buycotts and, 198–200, *199, 201,*
208–211
certification programs and, 208, 210
civic participation and, 205, *205*
effectiveness of, 211–213
electoral participation and, 204–205, *205*
ENERGY STAR program and, 209–210,
211–213
online participation and, 205, *205*
organization of political consumers and,
205–208
political consumerism described and,
198–200, *199*
political consumer types and, 200–205,
201, 202, 203, 204
Toyota Prius commercial hybrid
gas-electric car and, 208–211, 213
Cool Cities program of Sierra Club, 103
COP (Conference of the Parties),
5, 6, 8, 149
COP-7 in Marrakesh, 6
COP-13 in Bali, 8
COP-15 in Copenhagen, 8–9, 13, 20–21
COP-16 in Cancun, 13
COP-17 in 2017 in Durban, 10
COP-17 in Durban, 10
COP-18 in Doha, 10
COP-19 in Warsaw, 10
Copenhagen Accord, 9–10, 21, 52n13
Courts, and national climate policy making
overview of, 23, 33, 39–41, *40,* 50
American Electric Power Co. v. Connecticut
in 2009 and, 42
Clean Air Act and, *40,* 42–45
Connecticut v. American Electric Power Co.
in 2005 and, 42, 43
environmental assessment and, 46–47
EPA and, 42–49, 52n18, 54n54, 54n66

executive branch policy making and,
33, 43–46, 50
federal regulation implications and,
47–49
*Massachusetts v. Environmental Protection
Agency* in 2007 and, 42, 43, 48,
55, 60
public nuisance litigation and,
41–43, 52n42, 53n42
Crocker, David, 231–232
Cuccinelli, Mark, 65

Daley, Richard, 98
Data on climate change
CO_2 emissions and, 1, 6–8, *7,*
28n1, 35–36, *36*
GHG and, 73–74
global climate politics and, 1–2, 28n1
international climate regime and,
6–8, *7*
Defender of Wildlife, 206
De Figueiredo John M., 175
DOE (Chicago Department of
Environment), 98
Donner, Simon D., 127–128
Donovan, Todd, 177, 181, 183
Druckman, James N., 129
Dunlap, Riley E., 126

Earth First!, 155
Ecological modernization, national climate
change movement in US, 153
Economic conditions
environment-economy conflict reduction
and, 133–135
national climate policy making and,
73–74, 77–78
public opinion on climate change in
context of, 126–127, 133–135
states' climate policies and, 77–78
ECOS (Environmental Council of the
States), 76–77
EDF (Environmental Defense Fund),
152, 154, *163,* 163–165
EERS (energy efficiency resource standard),
67–68, *68–69,* 70

Electoral participation, and consumer
  political action, 204–205, *205. See also*
  Ballot measures, and states' policy
  making
Emissions management, and politics of
  transition, 221–224, 237nn13–14
EMOs (environmental movement
  organizations), 149, 151–152,
  154–155, 161–162
Endangered Species Act, 34, 46, 50
Energy Action Coalition, *151,* 155,
  159, 160, *162*
Energy efficiency resource standard (EERS),
  67–68, *68–69,* 70
Energy Security Act in 1980, 52n18
ENERGY STAR program and,
  209–210, 211–213
Environmental assessment, and courts in
  context of national climate policy
  making, 46–47
Environmental causes beliefs, and
  public opinion on climate change,
  112–113, *114*
Environmental Change Institute, 207
Environmental concern for climate
  change, and public opinion,
  113, 113–114, *115*
Environmental consciousness, and
  public opinion on climate change,
  110, 112, 119, 125
Environmental Council of the States
  (ECOS), 76–77
Environmental Defense Fund (EDF),
  152, 154, *163,* 163–165
Environmental movement organizations
  (EMOs), 149, 151–152, 154–155,
  161–162
Environmental Protection Agency (EPA),
  42–49, 52n18, 54n54, 54n66, 60–64,
  73–74, 77. *See also specific court cases*
Environment-economy conflict reduction,
  and public opinion on climate change,
  133–135
EPA (Environmental Protection Agency),
  42–49, 52n18, 54n54, 54n66, 60–64,
  73–74, 77. *See also specific court cases*

EU (European Union), 6, 10, 19–20,
  72, 133
EU-15 (15 EU member states
  1995–2004), 6
EU-27 (27 EU member states
  2007–2013), *36*
European Union (EU), 6, 10,
  19–20, 72, 133
European Union Emissions Trading
  Scheme, 72
European Union Emissions Trading
  Scheme, 72
Executive branch policy making, and
  courts, 33, 43–46, 50. *See also* Federal
  legislation climate policy making;
  National (federal government) climate
  policy making
Explicit climate actions, and cities' climate
  policy making, 91–92, *92*
ExxonMobil, 198, 206, 212

Federal legislation climate policy making.
  See also Executive branch policy
  making; National (federal government)
  climate policy making; States' climate
  policies
  overview of, 34, 36–39, *38,* 42–44,
    47–49, 52n18, 53n42, 57–59
  congressional hearings on climate change
    and, 148–149, 151
Federal Trade Commission's Energy Guide
  label, 210
ForestEthics, 211–213
Fossil fuel rejection, and delegitimization
  in context of politics of transition,
  224–228
Fovell, Robert G., 121
"Fracking" techniques (hydraulic
  fracturing), and states' climate policies,
  74–75
Frames or discursive frames for climate
  change, and national climate change
  movement in US
  overview of, 152
  global level climate change discourse and,
    152–154

national level climate change discourse and, 154–158, *159*, 160
resource mobilization level of climate change discourse, 155–157
Framing public policy, and public opinion on climate change, 130–133, 134
Freudenburg, W. R., 125
Friedman, Monroe, 213
Friends of the Earth, 163, *163*
Fuel efficiency for automobiles (vehicular fuel efficiency), 62, 75, 119, 210

G-8 countries, 8
G-77 (Group of 77 [developing countries]) +China, 11, 20–22
Gamson, W. A., 160, 165
Garrett, Elizabeth, 175
GCC (Global Climate Coalition), 112, 134, 212
Generational change (age cohort differences), and public opinion on climate change, 116–118, *117, 118*
Gerlak, Andrea K., 175, 176, 187
German Advisory Council on Global Change, 227
Germany, *7*, 7–8
GHG (greenhouse gases). *See* Greenhouse gases (GHG)
Global Change Research Act in 1990, 37, 45
Global Climate Coalition (GCC), 112, 134, 212
Global climate politics. *See also* International climate policy making as collective action problem; International climate regime
overview of, 2–3, 21–22, 25
COP-15 Copenhagen and, 8–11
data on climate change and, 1–2, 28n1
GHG and, 1–2
Kyoto Protocol and, 5–8
legally binding agreements and, 2, 5, 10, 12, 28n6
Global Climate Protection Act in 1987, 52n18
Global Environment Facility, 12

Globalization, and politics of transition, 229–230
Global level climate change discourse, 152–154
Goffman, Erving, 152
Gore, Al, 5–6, 127, 149, 154–155, 164, 207
Government-focused initiatives in context of community emissions, and cities' climate policy making, 90–91
Graham, Lindsey, 38–39
Great Britain, 34
Greenberg, J., 130
Green-e, 210
Green governmentality, and national climate change movement in US, 152
Greenhouse gases (GHG)
overview of, 1–2
asymmetry and, 13–15
cities' climate policy making and, 82–83, 88–89, *89, 90,* 91–92, *92–93*
data on, 73–74
federal legislation climate policy making and, 53n42
G-77+China and, 11
global climate politics and, 1–2
international climate policy making as collective action problem and, 3–5
international climate regime and, 9
Kyoto Protocol and, 6, 8
legally binding agreements and, 5, 10, 12
multiple issue set in context of adaptation and, 12–13
negotiation blocks and, 20–22
See also CO_2 and Chlorofluorocarbons
Green Party (U.S.), The, 206
Greenpeace, *163*, 163–164, 206, 211
Green Seal program, 208, 210
Group of 77 [developing countries] (G-77) +China, 11, 20–22
Guber, Deborah Lynn, 173, 176, 177, 187
Guzy, Gary, 44

Hansen, James, 10, 111, 148
Hofmeister, John, 227
Home Depot, 206, 208, 212–213

Human activity (biophysical) context of energy, and politics of transition, 221

Hydraulic fracturing ("fracking" techniques), and states' climate policies, 74–75

ICLEI (International Council for Local Environmental Initiatives), 83, 85, 86, 87–88, 93, 103

Implicit climate actions, and cities' climate policy making, 91–92, 92

An Inconvenient Truth (film), 127, 149, 164

India, 6, 7, 8–9, 14–15, 20, 36, 36, 44, 56

Individualistic participation, and consumer political action, 205, 205

Information conditions, and international climate policy making as collective action problem, 16–19, 17

Institutional change, and national climate change movement in US, 147–148

Interfaith Climate Change Network, 151, 158, 159, 162

Interfaith Power and Light, 158

Intergovernmental collaboration on climate policy making, 76–77

Intergovernmental Panel on Climate Change (IPCC), 3–4, 5, 12, 14, 16–17, 17, 19, 22, 45, 111, 116

International Center for Technology Assessment, 44

International climate policy making as collective action problem. See also Global climate politics

overview of, 11–12

ARs and, 3, 28n8

asymmetry and, 13–15

CO$_2$ emissions, 14, 16, 18–19

COP-7 in Marrakesh, 6

COP-13 in Bali, 8

COP and, 5, 149

GHG and, 3–5

information conditions and, 16–19, 17

multiple issue set in context of adaptation and, 12–13, 29n24

negotiation blocks and, 19–21

International climate regime. See also Global climate politics

overview of, 3

CO$_2$ emissions, 1, 6–10, 28n1

COP-15 in Copenhagen and, 8–9, 13, 20–21

COP-16 in Cancun and, 13

COP-17 in Durban and, 10

COP-18 in Doha and, 10

COP-19 in Warsaw and, 10

Copenhagen Accord and, 9–10, 21, 52n13

data on climate change and, 6–8, 7

GHG and, 9

IPCC and, 3–4

Kyoto Protocol and, 5–8, 28n13

UNFCCC in 1992 and, 3–5, 11, 28n12, 85

International Council for Local Environmental Initiatives (ICLEI), 83, 85, 86, 87–88, 93, 103

International Energy Conservation Code, 67–68

Internationally negotiated climate commitments, and national climate policy making, 33–35, 51n4, 52nn13–14

Interwest Energy Alliance, 151, 156, 158, 159, 160, 162

IPCC (Intergovernmental Panel on Climate Change), 3–4, 5, 12, 14, 16–17, 17, 19, 22, 45, 111, 116

Jenkins, J. Craig, 126

Ji, Chang Ho, 175

Johnson, Sadu, 98

Joint implementation (JI), 6

Jones, Charles, 135

Kansas, 69, 71

Kansas City, Missouri (KCMO) climate protection movement, 93–94

KCMO (Kansas City, Missouri) climate protection movement, 93–94

Kerry, John, 38–39

Kinsel, Sheldon, 146

Knowledge about climate change, and
public opinion on climate change,
114–116
Kousser, Thad, 175
Krosnick, Jon A., 132
Kunda, Z., 125, 128–130
Kyoto Protocol
overview of, 1–2, 5–8, 28n5
cities' climate policy making and, 86
GHG and, 6, 8
international climate regime and,
5–8, 28n13
JI or joint implementation and, 6
national climate policy making and,
34–35, 37

Lake, Laura M., 174, 177, 179–180
Lanz, Bruno, 177
League of Conservation Voters,
37, 160, 207
Legally binding agreements, and global
climate politics, 2, 5, 10, 12, 28n6
Leopold, Aldo, 231
Levy, David, 135
Lieberman, Joseph, 37–39
Local, the, and politics of transition,
229–230
Local Agenda 21, 85
Local climate policies, and cities' climate
policy making, 88–94, 89, 90
Localization, and politics of transition
overview of, 228–229
ethics of, 231–233, 238n31
local, the versus, 229–230
politics of, 230–231
Locally led initiatives, and impacts on cities'
climate policy making, 100–102
Lövbrand, E., 152, 153, 155
Lowe's, 206, 208
Luntz, Frank, 132
Lutrin, Carl E., 176, 177
Lutsey, Nicholas, 100

Magleby, David B., 171, 174, 175
Maine, and Question 2 or Energy Efficiency
Bonds Issue, 187–188

Markey, Edward, 48, 58, 63, 74, 135
Massachusetts, 44–45, 61, 68, 68, 176
*Massachusetts v. Environmental
Protection Agency* in 2007, 42, 43,
48, 55, 60
Mayors' Climate Protection Agreement
(MCPA), 86, 86
McCain, John, 37–38, 127
McCright, Aaron M., 126
McDaniels, Jeremy, 127–128
McGuire, William J., 121
MCPA (Mayors' Climate Protection
Agreement), 86, 86
Media, and public opinion on climate
change, 111, 123–124
Meral, Gerald, 175
MGGRA (Midwestern Greenhouse Gas
Reduction Accord), 64, 76
Michigan, 68, 71, 72, 171, 188–190
Michigan Energy, Michigan Jobs,
171, 189
Midwestern Greenhouse Gas Reduction
Accord (MGGRA), 64, 76
Migratory Bird Treaty Act in 1916, 34
Mobilization for Climate Justice,
151, 155, 159, 162
Montreal Protocol in 1987, 34
Motivated reasoning in context of
information, and public opinion on
climate change, 128–130
Motivations, for cities' climate policy
making, 94–99, 95, 97
Multijurisdictional climate policy initiatives,
72–73
Multiple issue set in context of adaptation,
and international climate policy
making, 12–13, 29n24

NA 2050 (North America 2050:
A Partnership for Progress), 76
NAPAs (National Adaptation Programmes
of Action), 13
Natali, Susan M., 175, 176, 187
National Academy of Sciences, 45, 52n18
National Adaptation Programmes of Action
(NAPAs), 13

National climate change movement in US
overview of, 24, 146–147, 164–165
civic environmentalism coalitions and,
153–156, *162*
coalition formation and, 149, 151
congressional hearings on climate change
and, 148–149, 151
ecological modernization and, 153
frames or discursive frames for climate
change and, 152
global level climate change discourse
and, 152–154
green governmentality and, 152
historical development of, 148–149,
150, 151, *151*
institutional change and, 147–148
national level climate change discourse
and, 154–158, *159,* 160
network analysis of social movements
and, 157–158, *159,* 160
power in context of influence in,
160–161, *162, 163,* 163–164
radical resistance in civic
environmentalism and, 153–154
reformism or participatory
multilateralism in civic
environmentalism and, 154
resource mobilization level of
climate change discourse and,
155–157
strong ecological modernization and,
153, 156–157, 161, *162*
UNFCCC and, 148–149, 151
weak ecological modernization and,
153–157, *162*
National (federal government) climate
policy making. *See also* Executive
branch policy making
overview of, 23, 32–33, 50, 55, 60,
62–63, 73
Clean Air Act and, 23, 33–34,
43–46, 48–49, 60–62
CO$_2$ emissions and, 35–36, *36,*
52n18, 53n42
collapse of, 56–59
economic conditions and, 73–74, 77–78

federal legislation climate policy making
and, 34, 36–39, *38,* 42–44, 47–49,
52n18, 53n24, 53n42, 57–59
hydraulic fracturing or "fracking"
techniques and, 74
intergovernmental collaboration and,
76–77
internationally negotiated climate
commitments and, 33–35,
51n4, 52nn13–14
Kyoto Protocol and, 34–35, 37
public opinion on climate change
and, 75, *75, 76*
UNFCCC in 1992 and, 34
unilateral domestic climate policy
and, 35–39, *36,* 52n18, 52n24
vehicular fuel efficiency and, 62, 75,
119, 210
National Environmental Policy Act (NEPA),
39–40, 46–47, 50
National Highway Traffic Safety
Administration, 45–46, 47
National level climate change discourse,
154–158, *159,* 160
National Wildlife Federation, 146, 148,
160, 163, 207
Natural Resources Defense Council
(NRDC), 148, 160, *163,* 163–165,
206–207
Nature Conservancy, 156–157, *163*
Negotiation blocks, and international
climate policy making as collective
action problem, 19–21
NEPA (National Environmental Policy Act),
39–40, 46–47, 50
Network analysis of social movements,
and national climate change movement
in US, 157–158, *159,* 160
New Jersey, 64, *68,* 72
Newman, Benjamin J., 203
New York City climate protection program,
101–102
Nichols, Mary, 77
Nicholson, Stephen P., 177
Nickels, Greg, 86
Nixon, Rob, 226

NOAA (U. S. National Atmospheric and Oceanic Agency), 18–19
Nonpolitical consumers versus consumer political action, 198, 200, *202,* 202–205
North America 2050: A Partnership for Progress (NA 2050), 76
NRDC (Natural Resources Defense Council), 148, 160, *163,* 163–165, 206–207

Obama, Barack, 9, 25, 35, 38–39, 49, 52n13, 57–58, 60–63, 65–66
OECD (Organisation for Economic Co-operation and Development), 6, 14
Online participation, and consumer political action, 205, *205*
Opinion formation dynamics, and public opinion on climate change, 121–128, *123*
Oregon, 67, 68, *68,* 85, 179
Organisation for Economic Co-operation and Development (OECD), 6, 14
Organization of political consumers, 205–208

Participatory multilateralism (reformism), in civic environmentalism, 154
Partisanship differences, and public opinion on climate change, 117–118, *118*
PCL (Planning and Conservation League), 175
PCL/F (Planning and Conservation League Foundation), 175
Perry, Rick, 61, 64–65, 70–71
Pew Charitable Trusts, 163, *163*
Pew Research Center for the People and the Press, 112, 117, 119, 122, 126
Pickens, T. Boone, 186–187
Planning and Conservation League (PCL), 175
Planning and Conservation League Foundation (PCL/F), 175
PlaNYC initiative, 101
Policy preferences, and public opinion on climate change, 118–119

Political consumerism. *See also* Consumer political action, and climate change description of, 198–200, *199*
political consumer types and, 200–205, *201, 202, 203, 204*
Politics of transition, and climate making policy
overview of, 25, 218–220, 233–234
biophysical context of energy and, 221
emissions management and, 221–224, 237nn13–14
ethics of localization and, 231–233, 238n31
fossil fuel rejection and delegitimization in context of, 224–228
globalization and, 229–230
the local and, 229–230
localization and, 228–233
politics of localization and, 230–231
transition defined and, 219–220, 236n3
Portney, Kent E., 102
Power, in context of influence in national climate change movement in US, 160–161, *162, 163,* 163–164
Proenvironmentalism
failures in, 172–177
successes in, 172–181, *179, 180, 182,* 183–185, *184*
Public Citizen, 206
Public nuisance litigation, and courts in context of national climate policy making, 41–43, 52n42, 53n42
Public opinion on climate change
overview of, 24, 110–111, 135–137
actions in context of attitudes about climate change and, 120–121
age cohort differences or generational change and, 116–118, *117, 118*
ballot measures and, 172–173
Clean Air Act and, 111
CO_2 emissions and, 111–112, 119, 127
conflicting opinions on climate change and, 111–112
consensus on climate change and, 111–112

conservative movement against scientific consensus and, 112, 115–116, 125–126
economy's impact on, 126–127, 133–135
environmental causes beliefs and, 112–113, *114*
environmental concern for climate change and, *113,* 113–114, *115*
environmental consciousness and, 110, 112, 119, 125
environment-economy conflict reduction and, 133–135
framing public policy and, 130–133, 134
knowledge about climate change and, 114–116
media's role in, 111, 123–124
motivated reasoning in context of information and, 128–130
opinion formation dynamics and, 121–128, *123*
partisanship differences and, 117–118, *118*
policy preferences and, 118–119
RAS or receive-accept-sample model and, 121
weather pattern changes and, 126–128
WTP or willingness to pay and, 120–121, 136
Public opinion on governmental climate change policy, 75, *75, 76*
Pyszczynski, T., 130

Radical resistance in civic environmentalism, 153–154
Rainforest Action Network (RAN), 160, 205–208, 211–212
Reality Coalition, 207
Receive-accept-sample (RAS) model, 121
Reformism (participatory multilateralism), in civic environmentalism, 154
Regional Greenhouse Gas Initiative (RGGI), 19, 64, 72–73, 76
Regional Integrated Sciences and Assessments (RISA), 18–19
Renewable portfolio standards (RPS), 66, 67, *70,* 70–72, 75

Resource mobilization level of climate change discourse, 155–157
Resources for the Future, 74, 148, *163*
RGGI (Regional Greenhouse Gas Initiative), 19, 64, 72–73, 76
RISA (Regional Integrated Sciences and Assessments), 18–19
Rising Tide North America, *151,* 152, 154, *159, 162*
Rocky Mountain Climate Organization, *151,* 155, 158, *159, 162*
Ross, Michael, 227
RPS (renewable portfolio standards), 66, 67, *70,* 70–72, 75

Schulkin, Brent, 197
Schwarzenegger, Arnold, 67
Scientific Certification System, 208, 210
Scruggs, Lyle, 127
Seasons' End, 154–155, 158, *159, 162*
Settle, Allen K., 176, 177
Shell Oil, 206, 212, 225, 227
Sherman, B. R., 129
Sierra Club, 103, 160, *163,* 163–164, 176, 189, 206
Smith, Daniel A., 175
Snyder, Rick, 189
Sperling, Daniel, 100
Standard Oil, 225
States' climate policies. *See also* Ballot measures, and states' policy making; Federal legislation climate policy making; National (federal government) climate policy making
overview of, 23, 55–56, 63–65, 73
California as dominant among, 65–67
coal phaseout and, 71–72
economic conditions and, 77–78
EERS and, 67–68, *68–69,* 70
EPA and, 77
hydraulic fracturing or "fracking" techniques and, 74–75
multijurisdictional climate policy initiatives and, 72–73
public opinion on governmental climate change policy and, 75, *75, 76*

RGGI and, 19, 64, 72–73, 76
RPS and, 66, 67, *70*, 70–72, 75
WCI and, 63–64, 66, 67, 72, 76
Stratmann, Thomas, 175
Strømsnes, Kristin, 205
Strong ecological modernization, and national climate change movement in US, 153, 156–157, 161, *162*
Sustainable Energy Coalition, 149, *151*, 155–156, 158, *159*, 160, *162*, 186
Sustainability, 85, 95, 102, 119, 218, 231

TckTckTck, *151*, 155–156, *159*, 160, *162*
350.org/1Sky, *151*, 155–156, 160, *162*
Texas, 61, 64, *69*, 70–71, 74, 87, 100
Three-Regions Collaborative, 76
Toyota Prius commercial hybrid gas-electric car, and consumer political action, 208–211, 213
Transition, and politics of climate making policy, 219–220, 236n3

Unilateral domestic climate policy, and national climate policy making, 35–39, *36*, 52n18, 52n24
Union of Concerned Scientists, 160, *163*, 206
United Nations (UN), 20, 52n13, 87, 111
United Nations Conference on Environment and Development (UNCED, or Rio Earth Summit, or Earth Summit) in 1992, 4, 11, 85
United Nations Environment Programme, 3, 16
United Nations Framework Convention on Climate Change (UNFCCC) in 1992, 1–2, 3–5, 11, 28n12, 34, 85, 148–149, 151
US Climate Action Network, 149, *151*, 156–157, *159*, 160–161, *162*
US Climate Action Partnership, *151*, 154, 156, *159*, 160–161, *162*
US Conference of Mayors' Climate Protection center, 103
U.S. General Services Administration's Carbon Footprint tool, 210

U. S. National Atmospheric and Oceanic Agency (NOAA), 18–19
US Public Interest Research Group (US PIRG), 103
Urban C02 Reduction Project, 85
USCAN, 155

Vedlitz, Arnold, 129
Vehicular fuel efficiency (fuel efficiency for automobiles), 62, 75, 119, 210
Via Campesina, 149, *151*, 155, 157, *159*, *162*

Waxman, Henry, 48, 58, 63, 74, 135
WCI (Western Climate Initiative), 63–64, 66, 67, 72, 76
Weak ecological modernization, and national climate change movement in US, 153–157, *162*
Weather pattern changes, and public opinion on climate change, 126–128
Webb, Wellington, 85
Western Climate Initiative (WCI), 63–64, 66, 67, 72, 76
Willingness to pay (WTP), 120–121, 136
Wolinsky-Nahmias, Yael, 176–177, 183, 187
Wood, B. Dan, 129
Woolsey, Jim, 227
World Meteorological Organization (WMO), 3, 16
World politics of transition, and climate making policy. *See* Politics of transition, and climate making policy
World Resources Institute, 59, *163*
World Values Survey, 198
World Wildlife Federation, 163
WTP (willingness to pay), 120–121, 136

Yale Project on Climate Change Communication, 114–115
Yeager, David Scott, 132

Zaller, John R., 121